Melanin
Is the Name but
Copper
Is the source of all pigmentation and
Life.

Melanin
Is the Name but
Copper
Is the source of all pigmentation and
Life.

By DARIUS CLAY

Saa Me Cu Publishing Company

ISBN-13: 978-0-9916688-0-9

Book cover illustration: Saa Me Cu Publishing Company and
Vivid Moon Media Group LLC.

TABLE OF CONTENTS

Preface

In 2006, I attended a conference on melanin held at the Northwest Activity Center on Meyers Street in Detroit Michigan. As I sat in the audience, listening to the various scholars speak on the different things that melanin does, the only thing that kept coming to my mind was: Where does melanin actually come from?

I knew melanin was a pigment, but I wondered what the source is! I knew that everything, such as the planets, stars, flowers, animals, fruits, vegetable, and the creatures of the sea has pigment in it since pigment is in everything, I wondered what is the source of the pigment melanin and all of the other pigments. My curiosity led me to do some research on pigment. So I started to study some of the ancient Egyptian text and the Metu Neter language (hieroglyphs). I began to understand some of the teachings of our ancient ancestors when it came to the studies of RA (the study of the science of the sun) its function in the universe, and how it was one of the major sources of pigment. The teaching led me to the Dogon tribe in Mali West, Africa. After reading two books, "*Conversations with Ogotemmeli, an Introduction to Dogon religious ideas*" by Marcel Griaule and "*The Pale Fox*" by M. Griaule & G. Dieterle), these books changed my life! For the first time I was able to see the universe through the eyes of my ancestors.

The books defined copper as the source of life, starting with the spirit and ending at the final stage of all creation. The Dogon called copper and water the "Grate Nummo Spirits". I have heard others call the "Grate Nummo" an amphibious creature that is half man and half fish. They say he is a man from the top down to his waist and a fish from the waist down to his feet. This description is a myth. Also, there's no teaching from the Dogon Tribe that states that the Nummo is a creature from outer space or an extraterrestrial from another planet. Anyone who says this, either does not understand the Dogon teachings or does not want people to know that the Dogon's have such a high level of knowledge without the use of modern day technology. These false statements are insults to the intelligence of the Dogon Tribe and my ancestors.

How to read this book: You should read the whole book first then go back and cross reference the chapters with one another.

Copper Beautiful You Are

GOD has made you from the dress plate of copper, so beautiful that no human eye will see you naked.

Your beautiful nature is not meant for thy to see.

If one look at you long enough One will be blinded by your rays of beauty.

GOD revealed your soul to us through his breathe Tefnut-*Moisture.*

She is that guardian mist that GOD has kissed, to reveal your reflection in the form of a rainbow.

Copper is your skin, but the fire of life is what you're giving.

Your ray is known as Aton Ra and it is from that ray you produce the beautiful brown skin in the reflection of your sweet divine loving.

You gave color to the people of the world, you put the rich rhythm in their souls better known as *Melatonin*, Which is the most divine part of your naked soul.

It is what keeps us in harmony with the consciousness above, like the beautiful brown skin you produced with your sweet love.

Copper is the color of your skin, but the fire of life is what you will be forever giving.

You change from gold, which is the reflection of GOD's soul,

To brown, which is the color you chose for the people with your soul,

From that, to all the colors of the rainbow and in your final stage, you turn green representing all vegetation that we see.

Copper you are more precious than gold.

It's no mystery to me,

Why GOD made you the essences of his soul.

Chapter 1

The History of Copper

Copper probably had its greatest development in Egypt, as early as 5000 B.C. Artifacts, weapons and implements left in graves were used by the dead were of copper. Definite records have confirmed working copper mines on the Sinai Peninsula, by King Seneferu, at about 3800 B.C. The discovery of crucibles at these mines indicates that the art of extracting the metal included some refining. The art developed to the point where copper has been hammered into thin sheets. As a result of the process pipes and other objects were formed. (The person who hammered the copper was known as the smith or the fashioner.)

During this period, bronze made its first appearance as a rod found in the pyramid of Meduin, the date of origin generally being accepted as about 3700 B.C. bronze an alloy, of primarily copper and about 12% tin. Copper lies between nickel and zinc and is in sub-Group 1 (copper, silver, gold) in the periodic classification of the elements. It is the ninth of the ten transition elements occurring between calcium and gallium and is the first element in which the atom has a complete third shell of 18 electrons. The electron arrangement indicates that the natural valence of copper is 1 and that the properties of its compounds in which the oxidation state + 1 is cuprous, (which Is copper's Latin name) should resemble to a lesser degree those of gold. This is true since the cuprous salts resemble to a great degree those of silver, and some of the naturally occurring compounds of the two metals resemble each other and frequently occur together.

There are a wide variety of naturally occurring compounds that make of cuprous salts, (Minerals), There are twelve essential minerals to the creation of life, and they are as follow;

1. Potassium Phosphate
2. Sulphate of Sodium

3. Chloride of Potassium
4. Fluoride of Lime
5. Phosphate of Magnesium
6. Sulphate of Potassium
7. Sodium Phosphate
8. Lime Sulphate
9. Silica
10. Calcium Phosphate
11. Sodium Chloride
12. Iron Phosphate

Chapter 2

What is Copper

Hydrogen is the first and only element, everything is made from it. However life could not be created from hydrogen. The Almighty Creator extracted water and copper from hydrogen and the two compounds are known as the first twin births of the universe. Copper and water atomic Number, accordingly to our ancient ancestors is eight. They are one and the same. The Dogon Tribe of Mali West Africa calls water and copper, The Grate Nummo Spirits.

Keep this in mind. *Whenever we use the term copper we are using it indiscriminately with water. They were known to our ancestors as the divine pair. Water is a source and carrier of copper which can be seen in a rainbow. You can also see the copper in water in little tear drops. With the projection of light, you can see the colors.*

Copper is the only metal compound that has pigments that no other metal compound produces. So all of the colors, in this beautiful universe, come from copper. Melanin is just the term for the name of the color of the pigment. It is not the source.

Copper, also known as Cuprite, is red in color and has eight sides. Because of its eight sides, our ancestors gave it the atomic number eight. Copper and water are in everything. The compound is the divine seed of the universe. Copper diluted in water solution is hydrated Cu (H2O) and is blue in color. With ammonia it forms the deep blue complex Cu (NHs3). An alkaline solution of copper with sugar gives a red crystalline material. They impart bright green color to the flame of the Bunsen burner, and in diluted solutions produce a brown color when treated with potassium Ferrocyanide. So, if you mix another compound with copper as well as with heat, water, and cold you will get various colors.

Copper produces many different salt crystals. These salt crystals are not just the source of all life, they are the compound of metal and nonmetal joined together by an ionic bond. A salt is made whenever an acid reacts with a metal or a base (a base is an alkaline), and they usually form beautiful color crystals.

The salt families in the metal (copper) comes from the base (copper oxide), and the nonmetal part (sulfate) comes from the acid (sulfuric acid).

Each acid, therefore, has its own family of salts. Sulfuric acid makes sulfates, citric acid makes citrates and so on. Each base also forms a family of salts. Copper oxide, for example, always makes copper salts.

Other Facts about Copper:

Copper is the source of Light: In the light spectrum the seven color rays create the white light we see.

Natural salts. Many ores and minerals are made of salts. (This is important to remember because all of these compound are a source of copper compounds.) These include limestone, calcium carbonate, which make up your bones. Gypsum (calcium sulphate). If they grow in the right conditions, all salts will form beautiful crystals colors.

Copper as a Transitional Metal. Metals that are compounded with copper sometimes form brightly colored compounds. Copper (II) Hydroxide can be prepared by adding a strong alkali to copper (II) salts which will give you a light powder blue color. Copper (II) Nitrate is hygroscopic, which means it absorbs water from the air; its color is blue. Copper Chromium compounds, (III) oxide, Cr2O3 is dark green in color. Copper Chromium (VI) oxide, CrO3 is red in color. Copper Lead (II) oxide is called litharge; its color is yellow or reddish. Copper (II) Carbonate contains the carbonate radical, CO2/3;

its color is light green. Copper (II) chloride contains water of crystallization and its color is blue-green.

Water. Water atomic number is eight, the same as copper, according to our ancient ancestors. Just like copper, water carries the pigment of copper. We can see these colors in water when we see the rainbow.

Copper is a Water-Soluble. Since alkalis break up in water to form ions, alkalis are good conductors of electricity. In an alkaline battery, the strong alkali potassium hydroxide is used to conduct electricity between two electrodes.

Copper Pigment Colors. Carotene is an orange color pigment, Xanthophyll is a yellow color pigment found in the blood of some insects, phaeophytin (a') is a gray-brown pigment, and Phaeophytin (b') is yellow-brown in its color.

Chlorophyll (a) is blue-green in its color, and Chlorophyll (b) is yellow-green in its pigment colors. Fucoxanthin is a brown pigment which would be in the melanin family. There are many more copper color pigments. These are just a few of them.

Copper is also an Alkaline. Alkalis are bases that can cancel out or neutralize acidity. All alkalis dissolve in water to form hydroxide ions (OH).These ions react with the hydrogen ions (H+) in acids to cancel out the acidity.

Periodic Table: *Copper is a Trace Elements* Trace elements are a form of copper, meaning copper is compounded with different minerals. By definition, trace elements are chemical compounds that naturally occurs in the soil, plants, and wildlife in minute concentrations. Also known as trace minerals, they are necessary for the optimal development and metabolic functioning of all living things. The alkaline earth elements are metallic elements found in the second group of the periodic table. All alkaline earth elements have an oxidation number of +2, making them very reactive. Because of their reactivity, the alkaline metals are not found free in nature.

Some of the Alkaline Earth Metals. Beryllium, Magnesium, Calcium, Strontium, Barium and Radium.

Copper is the source of alkali metals. Found in group 1 of the periodic table (formerly known as group IA), are very reactive metals that do not occur freely in nature. These metals have only one electron in their outer shell. Therefore, they are ready to lose that one electron in ionic bonding with other elements. As with all metals, the alkali metals are malleable, ductile, and are good conductors of heat and electricity. The alkali metals are softer than most other metals. Cesium and francium are the most reactive elements in this group. Alkali metals can explode if they are exposed to water.

Some Alkali Metals are. Lithium, Sodium. Potassium, Rubidium, Cesium and Francium.

Chapter 3

Pigmentation Comes From Copper

Copper and water are in everything: The metal copper is the only source from which all pigmentation comes. There are many trace minerals that have copper combined within them. I don't want to get too deep into this chapter when it comes to pigment. However I'm going to mainly focus on the minerals, because all life is made from minerals, including plants, insects, fish, birds, animals and humans. Most minerals have traces of copper in them, this is where their colors come from. In life all minerals take their shapes and forms from the vibrating frequencies that are being projected from the planets and the stars in the universe.

There are at least 224 types of minerals, I will not mention them all, but will discuss a few. As stated before copper and water are one and the same source, they are indiscriminate. Since copper mutates or transforms, the Creator uses water as the source to outline the design of all things. Before the creational process takes place, water is programmed and used to shape the image of all things; copper is the physical image of all things. The metal copper is the material source of all things as they take on the geometric forms as they occupy space. Copper is not just the source of what is taking shape, but it also gives us the geometric design in the patterns of colors.

Minerals: Minerals are substances that form naturally in the earth, they are different from rocks. They are compounds of other minerals with traces of copper in them, (Copper is what give minerals there colors) minerals form in many ways. The mineral halite, which is used as a table salt forms when water evaporates in a hot shallow part of the ocean, leaving behind the salt it contained. Many types of minerals are made of molten rock or magma, which cools and turns into a solid. Talc a mineral that can be used to make baby powder is formed deep in the earth, as the high pressure and temperature cause it to change into a solid rock. The colors of minerals are affected by a number of things such as temperature, atmospheric conditions, water, other chemicals, minerals, and light, etc...

Cadmiums: Cadmium is a chemical compound, it is a soft bluish-white metal which is chemically similar to the two other stable metals, zinc and mercury. Cadmium occurs as a minor component in most zinc ores and therefore is a byproduct of zinc production. It was used for a long time as a pigment and for corrosion resistance plating on steel while cadmium compounds were used to stabilize plastic, due to its high concentration of copper, cadmium is sometimes used in fertilizers. It is also used in batteries, nuclear fusion, black and white television phosphors, and blue and green phosphors for color television picture tubes. In paint, cadmium forms various salts, this sulfide is used as yellow pigment. Cadmium selenide can be used as red pigment, commonly called cadmium red.

Cobalt: Cobalt is a chemical compound and its atomic number is 27, like nickel. Colbert in the Earth's crust is found only in chemically combined form. Cobalt based blue pigments (known as Colbert blue) has been used since ancient times for jewelry, paints and to impart a distinctive blue tint into glass, but the color was later thought by alchemists to be the metal bismuth.

Cobalt is the active center of coenzymes called Cobalamins. The most common example is vitamin B12. As such, it is an essential trace dietary mineral for all animals. (There really is no such thing as vitamins, so called vitamins are only minerals combined with copper.) Cobalt inorganic form is also an active nutrient for bacteria, algae and fungi. There are different color cobalt such as: green Cobalt, black cobalt, blue cobalt and cobalt (II) chloride hexahydrate, which is red in color.

Ochre: Ochre ranges in color from yellow to deep orange or brown. It is also the name of the colors produced by this pigment, especially a light brownish yellow. Variant of ochre contains a large amount of hematite, or dehydrated iron oxide. It has the reddish tint known as red ochre. Ochre is a family earth pigments which includes yellow, red, purple, sienna and umber. The major ingredient in all the ochre is iron (III) oxide, known as limonite, which gives them a yellow color.

Stolzite: Stolzite is a mineral of lead and tungsten. Tuscan is a strong, heavy metal with an extremely high melting point (3370degrees Celsius). It is use as a filament light bulbs, the electric contacts in spark plugs and as electron targets in x-ray tubes. Stolzite has some black, orange and brown colors

Wolframite: Wolframite is a mineral of Tuscan with iron and magnesium. Its color is sometimes brown and black.

Aeschynite: Aeschynite contains yttrium, thorium, titanium, and niobium along with iron and calcium. It is a mixture of black and brown color.

Chromite: Chromite is an oxide of chromium and iron. Cranium is a silvery white metal with a blue tingle. Its strength, corrosion resistance and high melting point is about 1830 Celsius. It has led to applications in coating other metallic objects. Chromite is black in color.

Crocoite: Crocoite is a mineral of lead and chromium (lead chromate). Crocoite's color, always orange or reddish-orange, comes from chromium, a valuable mineral.

Hollandite: Hollandite is a mineral of manganese with barium and potassium. It is black in color.

Jeremejevite: Jeremejevite is a mineral of aluminum with fluorine and boron (a borate). It is light blue in color

Montroydite: Montroydite is a mineral of mercury it is gray blue and black in color.

Sapphire: Sapphire is a type of mineral Corundum. In fact, any gem quality corundum is called sapphire unless it is deep red, and then it is called ruby. A trace of chromium creates pink, Iron creates green and yellow, chromium plus Iron creates orange, titanium plus iron creates blue.

Taafeite: Taafeite is a mineral of beryllium, magnesium and aluminum. The principle ore of beryllium is beryl. Gem varieties of beryl include emerald, which is beryl with traces of chromium to contribute to the green color.

Wulfenite: Wulfenite is a mineral of lead and molybdenum. Its color are orange, brown and yellow.

Umber: Umber is a natural brown or reddish- brown earth pigment that contains iron oxide and manganese oxide. It is darker than the other similar earth pigments, ochre and sienna.

Plant Life. When it comes to plant life, there are other types of pigments Chlorophyll A absorbs red and blue-violet but reflects mostly green, giving plants with the chlorophyll A an intense green color. Chlorophyll B varies slightly, reflecting green and yellow light. Carotenoids absorb blue light and reflect red, orange and yellow. Anthocyanin, which absorbs blue and green light, gives some flowers and fruits, like blueberries and cherries, a purple or red color. Phycobilin absorbs orange, yellow, and green light and often is found in algae and Cyanobacteria. As you can see, there are many difference types of pigments depending on what it is you are describing. In the human chain, melanin is the term. In the plant chain, Chlorophyll A, Chlorophyll B, Carotenoids (where carrots get their name), Anthocyanin, and Cyanobacteria are the terms.

Chlorophyll. According to our ancient ancestors, previously our blood was green, the color of chlorophyll. Chlorophyll is copper at its pure oxidation state. (Oxidation is defined as the interaction between oxygen molecules and the different substances they may contact, from metal to living tissues.) With the help of oxygen, copper produces salt crystals and changes to a green color. When the tree leaves changes their colors from green to the various colors we see in the fall, it does this with the help of temperature change, water, sunlight, and sugar. It is the sugar that causes the copper to change from green, yellow, to a red color. The salt crystals are alkaline base.

These alkaline molecules rotate in the direction of the universe and acid molecules rotates in the opposite direction. This is important to know because this process of alkaline rotation in one direction and acid in the other direction, is the cause of fruits and vegetation to change their color and their sugar content. As the molecules move from their original alkaline state, which is green, they begin to move in the reverse order and become acidic. They begin to ripen or rotten. This process is caused by the sugar content which turns into an acid. So the sweeter the berry, in some cases, the more rotten the fruit.

Pigment In humans and animals. Copper and iron oxide is the cause of the color melanin. There are many types of iron, but I am going to name three of the main ones. Iron (III) oxide, known as ferric oxide, is the organic compound with the formula Fe2o3. It is one of the three main oxides of iron. The other two are iron (II) oxide (Feo), which is rare, and iron (II, III) oxide (Fe3o4), which also occurs naturally as the mineral magnetite.

Hematite: Hematite is a mineral known as (Fe2o3) and also as ferromagnetic dark red. It is readily attacked by acids. Iron (III) oxide is often called rust, and to some extent has this label because rust shares several properties and have a similar composition. This composition of copper, iron and hematite in the human blood causes the color pigment for melanin. Hemoglobin is derived from the name hematite.

Limonite: Limonite is an iron ore consisting of a mixture of hydrated iron (III) oxide-hydroxides in varying composition. The generic formula is frequently written as Fe0 (OH0). In its brown form, it is sometimes called *brown hematite* or *brown iron ore*. In its bright yellow form, it is sometimes called *lemon rock or yellow iron ore.*

Copper is the source of melanin and all pigments. Melanin is just a term that is being used to determine the color of the pigment and not to say it is the source of pigmentation. The pigment melanin is a class of compounds, these compounds are copper. (Hematite pigment is the cause for brown to black skin complexion and limonite is the pigment that causes light skin to yellow complexion in people.) They serve as a pigment with different structures responsible for dark and tan pigments. There are different classes of pigments: black and brown insoluble eumelanins pigments, which are derived from aerobic oxidation of tyrosine in the presence of tyrosine, and the alkali-soluble phaeomelanins, which through the intervention of red -brown colors, arising from the deviation of the eumelanins pathway through the intervention of cysteine and or glutathione. Manganese oxide will give you a blue black pigment when mixed with copper. The region in which you live, the climate of that region, the foods you eat, and your amount of exposure to the sun will determine the pigment of your skin.

Chapter 4

Copper is the Spirit Light which is the cause of the Electrical Principle, and Magnetic Energy

Copper is projected from the sun in the form of rays. The rays that are projected move in the form of water. The rays are absorbed into the skin of all life forms. Although our eyes can only see the white light from the sun, that is not the only light that is being projected from the sun; there are seven different colors being projected as well. These colors are known as the ultraviolet light. When the ultraviolet light comes together, they form the white light that we see.

Water and copper are one in the same source. They both come together to form the spiritual aspects of life. The spiritual aspects of copper and water would be difficult to explain without first explaining the principle aspects of their nature. The first principle aspect is copper, in the form of pure energy. The second one is copper and water in the form of matter, they come together to form the activity of the quickening spirit. Pure energy is direct current undifferentiated spiritual principle. It is pure qualitative, it is the energy of divine life and inconceivable. It is the principle of life in action, the very essence of form, and the cause of all forms.

Copper and water are the principle of substance which is the source of all life. It is the essential fire source of form. Its work is the divine source, not created. It is abstract spiritual, like the cosmic egg fire that quickens it. Pure energy, copper and water in the abstract, are attributes of the cosmic virgin. Some say it is the mystery that can never be put into human language.

The perpetual virgin, united with the divine word (fire) that fertilizes her and which she bears eternally, is the mother of that uncreated light that our ancient ancestors called RA (the Sun). So out of the cosmic egg, which is made from pure energy, copper, and water we get the first born, the (Sun). The Sun is the source of light in the universe.

Pure energy, copper and water are the form of continuous creation. The three come together to form the alternating energy which is the cause of the magnetic field that gives aspect to everything in time, lightness, darkness, warmth, coldness, expansion, and contraction. It is the source that cause the duality in nature.

The magnetic field is also known, according to the Dogon tribe of Mali, West Africa, as the spider which produced the first word; not the first spoken word by God which was the first sound of the making of the universe. No, this word formed the fiber skirt which is the magnetic skin or aura that surrounds everything; it produces its own frequency. The earth has its language and its own vibrating frequencies known as the Bayuali; by the Dogon tribe. All frequencies travel along the magnetic field which includes, radio waves, micro waves, ultra-waves, gamma waves, and all vibrating frequencies of the planets, and even the words we speak.

The sun is one of the major causes of the magnetic energy which is made from copper and water. This field also protects the earth, as well as keeps everything in its place. This magnetic field also causes the rotating motion of the planets; this rotating motion is just like a generator.

In its simplest form, pure energy is the spirit energy, fire unseen. Whereas copper and water are the fire substance of matter, which is the divine seed. So in order to transform the pure spirit energy into a source that can be seen in the form of light, it needs copper and water as its source.

Like the sun in its rays, the spirit is a reflection of the sun light and just like the sun light, the spirit light projects the seven color rays.

Magnetism. Let me first say that you cannot have magnetism without electric currents. Copper is the source of the electric field that produces the currents that generate the magnetic energy, although copper is not effected by magnetism.

They say that magnetism is a physical phenomenon that includes forces exerted by magnets on other magnets. (This is why gravity exists.) Everything is affected including the earth, humans, animals, plant life, all life, planets, and the sun. Nothing is exempt from the magnetic forces; it is how everything is being held together. This force acts as a glue. It has its origin in electric currents and the fundamental magnetic moments of elementary particles. These give rise to a magnetic field that acts on other currents and moments. All materials are influenced, to some extent, by a magnetic field. The strongest effect is on permanent magnets, which have persistent magnetic moments caused by ferromagnetism, the basic mechanism by which certain materials, such as iron; which forms permanent magnets. Most materials do not have permanent moments, some are attracted to a magnetic field (paramagnetism); others are repulsed by a magnetic fields (diamagnetism); and others have a much more complex relationship with an applied magnetic field (spin glass behavior and antiferromagnetism).

Substances that are negligibly affected by magnetic fields are known as non-magnetic substances. They include copper, aluminum, gases, and plastic. Pure oxygen exhibits magnetic properties when cooled to a liquid state.

The Magnetic Field and the Universe. According to our ancient ancestors, our galaxy began in the center from the star Sirius, which is known as the placenta of our galaxy. This placenta was broken up to create the smaller stars in our galaxy. Our very own sun was given birth from the placenta of the star Sirius. Then the sun gave birth to our planet, Earth, and all the other planets in our solar system. Planets rotates on their own axis like stars. Satellites, such as our moon, are a piece of earth which does not rotate on an axis. All planets have a core within their center, this core is a small sun, a piece of our suns placenta which is made of copper. The core rotates, acting as a generator, and produces a magnetic field. The planets magnetic field varies on different timescales, but the most important variation is the one that results in field polarity reversals. The dynamo process behaves like a magnetic induction coil that allows energy to be produced by electromagnetic induction. When the earth's outer core, which is made of iron and nickel, spins around the center core, the earth's molten layer, it causes a current to flow around the outer layer. As the molten metal (copper) goes through the stronger fields, the fields vary according to the outside forces; the sun, other planets, other stars, etc.... More current is induced that increases the field even further.

This geomagnetic dynamo process needs energy to keep running. It takes millions of tons of energy to produce and sustain the magnetic field inside the earth. Thus, the magnetic field cannot exist at all if the dynamo does not depend on some external supply. The system cannot sustain itself by using fuel from the earth core. This current is what powers the earth's internal magnetic field and causes the earth to spin. Basically, the inner core rotates because it is part of a vast electric circuit connected to the sun.

However the energy supply from the sun can remain more or less steady for a very long time, but incremental changes will always take place, on upward or downward spiral cycles. The upward and downward cycles are when the planets move up and down, like bouncing balls, in reference to the sun, which is also the cause of the seasonal changes. The seasonal changes are not caused by the tilting of the earth on its axes, according to modern day scientist. Change during the upward cycle can be further enhanced if an excess amount of energy from the sun is thrown towards the earth, or if there is an increase in interstellar gamma radiation, or cosmic ray partials (color rays projected by all light). These would induce extra magnetized plasma in the outer layers and transfer it to the inner layers. This is why the core temperature of the earth depends on the total amount of extraterrestrial radiation.

There are other energy sources, like radioactivity and current, but the process of electromagnetic induction does not put out a constant current because it depends on external factors. One of the main factors is the star Sirius. It has a strong effect on earth's magnetic field, especially when our sun aligns with it every sixty years.

The impact of the sun's magnetic field on earth increases with or under three conditions. Those conditions are: when the earth gets closer to the sun; when the magnetic attraction is maximized (which apparently happens every 33 days and is perhaps caused by the moon); and finally during periods of high solar activity. A solar flare that is magnetized temporarily, strengthens the sun magnetic field and ionizes atoms on earth and instigates huge disturbances in their nuclei.

This strong effect causes the nuclei to be excited to higher energy levels, which in turn triggers their magnetic moment to change and results in faster nuclear decay rates for radioactive elements.

Charged particles move clockwise and the other particles move counterclockwise. Particles and subatomic particles can be coherently structured in only three possible ways. One type has a higher number of left hand rotations and is called negativity charged particles. The other type has a higher number of right hand rotations and is called positivity charged particles. The third kind is formed from an equal number of left handed and right handed charges, and this is called a neutral particle.

Properties of atom particles and subatomic particles are defined by the number of these charges, their orientation, and how they are structured to form a coherent object. The transfer of energy among these charges in their orbits is resonant and instantaneous. It is important to note that the force that governs the physical reality and everything else is only an appearance and a manifestation of the magnetic structure of matter.

Chapter 5

Copper Is the Fire of Life

In the sense of activity, copper is fire in the active state of life, it is connected to our senses through smell, taste, light and consciousness. This experience comes from the pigmentation of copper with the aid of water. Without water, we cannot experience taste or smell. It is through our eyes that we experience light which is active through water, light is the source of universal consciousness. Fire means activity or anything that is active. Not to be confused about the word "fire" and its properties but rather "fire" (which is pigmentation) is the driving force for all living things and their distinctive colors, shapes and sizes. It controls, and regulates the development and growth. Copper is the reason of activity in its electrical state, sometimes hidden or unseen. This is the state that drives the spiritual aspects of nature. In this state copper is the source of spiritual consciousness.

The Egg of a Chicken. Let us take a look at the egg of a chicken. The egg is made of the four basic elements: water, fire "copper", earth, and air. Clear liquid inside the egg is water in the passive state, which is copper, (remember water and copper are one and the same.) Therefore, water is a part of the placenta giving it a double meaning. Since the placenta is considered copper, that which is being birth from it is also copper. The shell of the egg is earth, and if you look closely into it you will see the air sack. At the blunt end of the egg between the two membranes, there are air pockets which are formed from the pores in the shell, this air cell provides oxygen for the developing bird during its incubation period. The yoke of the egg is the active state of fire that carries the pigment. The yolk color of the bird's egg solely depends upon the food eaten by the female. The shell's color is determine by the food eaten, as well. Egg yolks of wild birds vary from pale yellow to a dark red that is almost a maroon color. There are rare cases when a chicken produces a yolkless egg. This occurs when something prevents the yolk from entering the oviduct. In these rare cases a portion of the white gets enclosed in the shell, and the egg is usually smaller, about the size of a pigeon's normal egg. This fact should be no surprise, because copper is the cause of our development and growth. The fetus feeds from the placenta for its growth and development. Therefore, a lack of fire (copper), in the stage of the development will cause the growth to stunt. Even In the development of the chicken shell the more pigment the yolk has in it, the harder the shell will become. The darker the yolk, the more you see copper. So if a yolk is light yellow, it is a sign that it lacks copper, which means the bird is sick or unhealthy. In some cases, chickens may not have access to food sources, such as grass, corn, and other food high in copper pigment. In those cases, they are given additives in their feed to enhance the yolk color. Now some additives may trigger allergic reactions in some people. So if you have an allergic reaction to some eggs, the cause may be the additives. So be mindful of where you are buying your eggs.

The Sun. Now let's take a look at another fire. In this case, the sun is an active force which is made of copper. This is because the sun projects copper in the form of ultraviolet rays, made up of the color spectrum of light, according to their wavelengths, red being the longest, orange the next and then yellow, green, blue, indigo, violet is the shortest. The waves move in the form of water. This water wave can also be absorbed from all life. That is why we say that it is water because water is the only source that can be absorbed. The sun is a living organ that gives off its vital source, which is copper, to all life. Without the sun, it would be impossible for life to exist in the way we know it today. The planet earth is also a living organ, but in order for it to be as such, it must also have a sun of its own. Its sun is the core, made of copper, which lies in the center of the heart of the earth. The sun is like the heart of the solar system. It pulsates its ultraviolet rays of copper, giving all life the essence of life in the form of fire. Throughout the universe, this fire is the source of activity. In the winter solstice, when the sun has very little effect on life, the activity of life slows down and in some cases goes completely dormant. This state is called the masters grip, which means the sleep state of consciousness.

Blood. The blood is the vital source of life for the human body. It is made of copper and iron and the heart is the pumping station that helps its movement throughout the body. The sun also pulsate its vital source of life, copper, in the form of ultra-violet rays of colors. This is the blood of the universe. It is vital for all plant life. It is what makes the plants blossom. The life source is copper (pigment) the cause of all activity. It is the vital fluid that is in all life and the divine seed of all life.

In order to experience copper in its active state you must have water which is copper in its passive state. They both work together as universal twins.

Chapter 6

Copper is the Placenta of all life and the Universe

The universal placenta is known as the Nebula, a form of colorful gases that forms in the beginning stages of the galaxy formation. Keep in mind that the beautiful colors or pigments in the Nebula is made of copper. So the copper gases in the universal placenta is where the creation of the planets come from, and the planets take form with their own personality in the form of colors.

The Sun as the Placenta: The placenta is defined as an organ that connects the developing fetus to the uterine wall to allow nutrient uptake, waste elimination, and gas exchange. The sun is the placenta of our galaxy. It provides the vital nutrients to its fetus which would be all the creatures of life, through the process of photosynthesis. It eliminate waste or kills germs with its ultraviolet light. The gas exchange is oxygen and carbon dioxide.

The Human Placenta: All placentas are made from copper. In the human life, the placenta is made from the semen. The semen is what gives birth to the placenta and the placenta gives birth to the fetus. The fetus feeds from the placenta to create a human life. Copper is a mutating compound, meaning it is constantly changing. In its growth the child is constantly changing also. The placenta and the child is one and the same, meaning that the child is made from the placenta. Since the child is made from copper then, the child is copper, which makes the placenta double in its nature.

Placental Hormones. In addition to the role in transporting molecules between mother and fetus, the placenta is a major endocrine organ. It turns out that the placenta synthesizes a huge and diverse number of hormones and cytokines that have major influences on ovarian, uterine, mammary and fetal physiology, not to mention other endocrine systems of the mother. (The major source hormone that control the sex organs is copper).

Steroid Hormones: Sex steroid are the best known examples of placental hormones. Two major groups are produced by all mammals:

Progestins: Progestins are molecules that bind to the progesterone receptor. Progesterone itself is often called the hormone of pregnancy, because of the critical role it plays in supporting the endometrium and hence on survival of the conceptus.

The placenta of all mammals examined produce Progestins, although the quantity varies significantly. In some species (human females, horses, sheep, cats), sufficient progestin is secreted by the placenta. The ovaries or corpora lutea can be removed after establishment of the placentas and the pregnancy will continue. In other animals (cattle, pigs, goats, dogs), luteal progesterone is necessary throughout gestation because the placenta does not produce sufficient amounts.

Progestins, including progesterone, have two major roles during pregnancy: One role of the endometrium, is providing an environment conductive to the fetal survival. If the endometrium is deprived of progestin, the pregnancy will inevitably be terminated.

Suppression of contractility in uterine smooth muscle, which, if unchecked, would clearly be a disaster. This is often called the "progesterone block" on the myometrium. Toward the end of gestation, this myometrial-quieting effect is antagonized by rising levels of estrogens, thereby facilitating parturition.

Progesterone and other Progestins also potently inhibit secretion of the pituitary gonadotropins luteinizing hormone and follicle stimulating hormone. This effect almost always prevents ovulation from occurring during pregnancy.

Estrogens: The placenta produces several distinct estrogens.

In woman, the major estrogen produced by the placenta is estriol, and the" equine placenta" synthesizes a unique group of estrogens not seen in other animals. Depending on the species, placental estrogens are derived from either fetal androgens, placental Progestins, or other steroid precursors. With few exceptions, the concentration of estrogens in material blood rises to maximal toward the end of gestation.

Two of the principle effects of placental estrogens are: Stimulated growth of the myometrium and antagonize the myometrial-estrogen in late gestation induces myometrial oxytocin receptors, thereby preparing the uterus for parturition.

Estrogens are one in a battery of hormones necessary for both ducal and alveolar growth in the mammary gland. Like progestins, estrogens suppress gonadotropin secretion from the pituitary gland. In species like humans and horses, where placental estrogens are synthesized from androgens produced by the fetus, maternal estrogen levels are often a useful indicator of fetal well-being.

Protein Hormones: Several protein and peptide hormone are synthesized in placenta of various species. They have effects on the mother's endocrine system, fetal metabolism and preparation of the mother for postpartum support of her offspring.

Chorionic gonadotropins: As the name implies, these hormones have the effect of stimulating the gonads, similar to the "pituitary gonadotropins." The only species known to produce a placental gonadotropin are primates and equids.

The human hormone is called human chorionic gonadotropin or simply hCG. This hormone is produced by *fetal trophoblast cells. It binds to the luteinizing hormone* receptor on cells of the corpus luteum, which prevents luteal regression. Thus, hCG serves as the signal for "maternal recognition of pregnancy." The first hormone you produced was hCG.

Equine chorionic gonadotropin is also produced by fetal trophoblast cells. It is actually the same molecule as equine luteinizing hormone.

Placental Lactogens: These hormones are molecular relatives of prolactin and growth hormones. These hormones have been identified in primates, ruminants and rodents, but not in other species.

The functions of placental Lactogens are not well understood. They are thought to modulate fetal and maternal metabolism, perhaps mobilizing energy substrates for fetal use. In some species they have been shown to stimulate function of the corpus luteum, and to participate in development of the mammary gland prior to parturition.

Relaxin: Relaxin is a hormone thought to act synergistically with progesterone to maintain pregnancy. It also causes relaxation of pelvic ligaments at the end of gestation and may therefore aid in parturition. In some of the species in which relaxin is known to be produced, it is produced by the placenta, while in others, the major source is the corpus luteum. In some species, relaxin is produced by both the corpus luteum and placenta.

The Earth Soil: The earth soil is the placenta for plant life. No growth can take place without copper being in the soil. Copper is the source of the devolvement and growth for all life, but the truth of the matter is that no seed can germinate itself without the copper in the soil.

Copper levels also drop as the amount of organic matter is increased. Organic matter usually hampers the availability of copper by reducing soil mineral fixation and leaching. However, once organic matter has sufficiently decomposed, adequate copper can be released into the soil and taken up by plants. When copper is added to the soil there are a number of factors that will affect its availability. Soils with a high organic matter and clay content will hold on to copper more than sandy soils. In fact, soil with organic matter binds copper more tightly than any other micronutrient. In these high organic matter high clay content soils, copper tends to accumulate on the surface soil and is unlikely to leach from the soil. As a result, the copper that actually is in the soil solution and is available for plant life is a very small percentage of what you started with. Inadequate levels of copper can lead to poor growth, delayed flowering, and plant sterility. Copper deficiency in plant growth may appear as wilting with leaf tips turning a bluish-green color. In grain-type plants, the tips may become brown and appear to mimic frost damage.

The trace minerals in the soil are known as macronutrients and micronutrients. These are the same minerals that are essential for human life, as well as plant life. These minerals react to plant life the same way they react to human life. Notice the pigment change in the plant when the copper content is not properly balanced.

Macro nutrients Nitrogen: Promotes leaf and stem growth. When the soil lacks sufficient nitrogen, growth is reduced and leaves turn a pale yellowish-green. When soil is cold and wet, nitrogen supplies in the soil aren't as available to the plants. Too much nitrogen in the soil can lead to a potassium deficiency.

Phosphorus: This nutrient is critical in the germination and growth of seeds, flowers, fruit, and roots. When phosphorus is deficient, growth is reduced and leaves fall off prematurely. Plants lacking phosphorus produce dull, bluish-green leaves that turn a purplish or bronze color with brown edges. Just like nitrogen, too much phosphorus may lead to potassium deficiency.

Potassium: Potassium Promotes hearty, disease-resistance growth. Without adequate potassium, growth is stunted and leaves grow close together. Leaf tips and edges become brown and edges curl. Too much potassium in your soil can lead to calcium and magnesium deficiencies.

Calcium: Calcium is a vital component in cell walls of plants. Calcium is essential for root growth. Without adequate calcium, roots develop poorly and leaves become distorted and often turn brown.

Magnesium: This nutrient is necessary for chlorophyll production and plays a vital role in most enzyme reactions. If your soil lacks the proper amount of magnesium, the deficiency causes varying symptoms in different plants. Most often symptoms include things like yellowing leaves that may suddenly fall off without withering. Too much magnesium could cause a calcium deficiency.

Sulfur: Sulfur is a necessary nutrient for chlorophyll formation. Too little sulfur produces slow growth that generates small round leaves that are stiff and brittle. Leaves drop and tip buds die.

Micronutrients: *Iron* - This trace element is needed for chlorophyll formation and oxygen transfer. Without adequate iron, leaves yellow (but leaf veins remain green) too much lime in the soil can lead to an iron deficiency.

Manganese: Manganese works as a channel for a variety of enzymes and is essential for chlorophyll formation. Too little manganese causes a variety of indicators depending on the type of plant. The most common signs include yellowing leaves with green veins or grayish-white specks that appear on the leaves. Too much manganese may deplete iron in the soil and cause symptoms similar to those exhibited with a lack of manganese.

Boron: Boron works as a vehicle for the interchange of sugars, for reproduction, and for cell intake of water. Without adequate boron your plants will be distorted with hollow stems and malformed fruit. Other symptoms include scorched, curled and sometimes mottled leaves.

Zinc: Zinc is important for the production of proteins. This nutrient affects how big plants grow and whether or not they mature. Lack of zinc produces less fruit and brings about yellowing of leaves between veins, often accompanied by purple or dead spots with small, deformed leaves growing close together.

Copper: This is another nutrient important for the production of proteins and plays an important role in reproduction. If the plants don't get enough copper, they will display bluish-green leaves that might wither or never unfold. Lack of copper may also form rosettes on growing tips.

Molybdenum: This nutrient is vital to nitrate enzymes and the formation of root nodules in beans and peas. Without adequate supplies of molybdenum, leaves produce yellow mottling and dead spots and often growing tips are distorted or killed.

Chlorine: Lack of the right amount of chlorine in the soil might affect carbohydrate metabolism and photosynthesis. This deficiency leads to stubby roots and wilting.

Compost Soil: Compost is known as humus. Humus is like semen to the soil it is made of decaying matter which Is a composition of copper. All life is made from copper when you take a look at what is being decomposed it is matter with copper being its source.

Composting utilizes the natural processes of decay to convert organic materials such as leaves, grass and food scraps into a valuable humus-like material called compost. A compost pile or bin allows you to better maintain ideal levels of air, moisture, food and temperature so that decomposition proceeds much faster than normal. With a properly managed compost pile or bin your leaves, grass, and food scraps can be turned into an excellent compost in just a few months.

The process of "humification" can occur naturally in soil, or in the production of compost. The importance of chemically stable humus is thought by some to be the fertility it provides to soils in both a physical and chemical sense, though some agricultural experts put a greater focus on other features of it, such as its ability to suppress disease. It helps the soil retain moisture by increasing micro porosity, and encourages the formation of good soil structure. The incorporation of oxygen into large organic molecular assemblages generates many active, negatively charged sites that bind to positively charged ions (cations) of plant nutrients, making them more available to the plant by way of ion exchange. Humus allows soil organisms to feed and reproduce, and is often described as the "life-force" of the soil. Yet, it is difficult to define humus precisely; it is a highly complex substance, which is still not fully understood. Humus should be differentiated from decomposing organic matter in that the latter is rough-looking material, with the original plant remains still visible, whereas fully humified organic matter is uniform in appearance (a dark, spongy, jelly-like substance) and amorphous in structure, and may remain as such for millennia or more. It has no determinate shape, structure or character. However, humified organic matter, when examined under the microscope may reveal tiny plant, animal or microbial remains that have been mechanically, but not chemically, degraded. This suggests a fuzzy boundary between humus and organic matter. In most literature, humus is considered an integral part of soil organic matter.

Note all life is made from copper. When the physical body dies and return back to the earth, it becomes semen, fertilizing the soil forming the placenta for the seeds that will be planted.

Chapter 7

Copper and iron transport across the placenta: regulation and interactions.

Iron and copper are both essential micronutrients and are required for a wide variety of enzymatic and processes within the developing fetus. Transfer of both nutrients across the placenta is tightly regulated. For copper, uptake is a carrier-mediated process, but their localization and how they are regulated is only now being elucidated. Regulation of copper proteins appears to be different from that of iron, with localization of the protein, rather than changing levels being responsible for altering rates of transfer. This may not be true for all the proteins and genes involved in the delivery of copper, and again, there is much that remains to be clarified.

Yearling ewes were dosed with a tracer level of radio copper for blood-balance and subsequent placental-transfer studies, using maternal-fetal copper utilization criteria in comparative studies.

Over 7% of an oral copper dose was excreted by 30 hours; post dosing over 92% by 120 hours; post dosing, whereas, 5.2% of an intravenous dose was excreted by 24 hours. Copper levels in total products of conception increased 27 times between trimesters one and three, containing 13.81 mg total copper after 136 total days of gestation, and radio-copper were highest in liver and increased in all fetal tissues as gestation advanced. Copper concentration was higher in the older fetuses. Ewes retained an average of 0.81 mg copper per day. After 34 days the fetus contained 0.004 mg of copper; [placenta, 0.47 mg, and placental fluid, 0.032 milligrams. The 90 day fetus contained 2.84 mg copper; placenta 1.88 mg and placental fluids, 0.55 milligrams. The calculated daily amount of copper deposited in the total products of conception during the first, second, and third-trimester, ewes averaged 15ug, 85.1ug and 1.86ug, respectively. These data indicated a concentration barrier for copper between ewe and the fetus in this syndesmochorial type placenta.

The placenta contained 1.5mg copper; 3.9% copper fluids; 0.96mg copper and undetectable amounts of copper and 2.2% of the copper dose after 136 days gestation. High copper content in most newborn animals has suggested placenta transfer and storage before birth (Neuweiler, 1942; Cunningham, 1931; Pryor, 1964). However, little or no quantitative tissue studies on copper placental transfer have been reported for ruminant animals.

Fetuses, Infants, and Children. Copper is essential for the normal growth and development of human fetuses, infants, and children. The human fetus accumulates copper rapidly in its liver during the third trimester of pregnancy. At birth, a healthy infant has four times the concentration of copper than a full-grown adult. Human milk is relatively low in copper, and the newborn child storage of copper falls rapidly after birth. Supplying copper to the fast-growing body during the breast feeding period, ensure enough copper for adequate growth and development. These supplies are necessary to carry out such metabolic functions as cellular respiration, melanin pigment (which is copper) and connective tissue synthesis, iron metabolism, free radical defense, gene expression, and the normal functioning of the heart and immune systems in infants.

Most well-nourished children have adequate intakes of copper. Health-compromised children, including those who are premature, malnourished, have low birth weights, suffer from infections, and who experience rapid catch-up growth spurts, are at elevated risk for copper deficiencies.

Fortunately, diagnosis of copper deficiency in children is clear and reliable once the condition is suspected. Supplements under a physician's supervision usually facilitate a full recovery.

Chapter 8

Copper is the Cause for Our Development and Growth

Copper is distributed throughout our bodies from our hormonal glands. Each gland has a function as to where it is going to distribute the mineral copper. We have to understand that the placenta is made from copper and the fetus is feeding from the placenta in order to grow. Copper is the cause of growth from the beginning.

Copper deficiency during pregnancy results in early embryonic death and fetal structural abnormalities including skeletal, pulmonary and cardiovascular defects. During pregnancy, copper transporting compounds are expressed in the placenta.

Copper is an essential trace element, necessary for the survival of all living organisms. It is central to the function of many enzymes, including cytochrome oxidase, lysyl oxidase a copper-containing enzyme responsible for the maintenance of collagen and elastin in tissues. Its absence leads to aneurysm formation and rupture of large blood vessels and superoxide dismutase, forming an integral part of the enzymes active site.

The importance of copper for humans is illustrated by two inherited disorders of copper metabolism, Menkes and Wilson disease. Menkes disease is an X-linked recessive disorder, caused by mutations in the gene MNK; ATP7A (1-3). The reduced absorption of copper from intestinal cells and defective distribution around the body in this disease leads to reduction in the activity of important copper – dependent enzymes. Key features of Menkes disease include abnormal hair structure, progressive cerebral degeneration, vascular aneurisms and thrombosis. Wilson disease is an autosomal recessive disorder caused by a mutation in the Wilson gene, WND;

ATP7B resulting in toxic copper accumulation in the liver and brain. Patients with Wilson disease show a range of clinical conditions including chronic hepatitis, cirrhosis and neurological disturbances including movement.

That MNK is important for normal development is evidenced in babies with Menkes disease. Due to copper deficiency, Menkes babies are lethargic, have sparse hair and develop hypothermia and convulsions, leading to death within the first year of life. Evidence suggests that MNK is involved in the transport of copper to the fetus, as an increase in the placental copper content is observed in a mouse model of Menkes disease, the brindled mouse, and a direct block of placental transport has been demonstrated in this mouse model. The role of WND in early development is not clear, but as copper accumulates in the placenta during gestation in both Menkes and Wilson disease a role for both ATPases in the transfer of copper across the placenta to the fetus is likely.

Follicle Stimulating Hormone (FSH) and Luteinizing Hormone (LH) are called gonadotropins because the hormone stimulate the gonads-in male's testes and the ovaries in the female. They are not necessary for life, but are essential for reproduction. These two hormones are secreted from cells in the anterior pituitary called gonadotrophs. Most gonadotrophs secrete only LH or FSH, but some appear to secrete both hormones.

In both sexes, LH stimulates secretion of sex steroids from the gonads. In the testes, LH binds to receptors on leydig cells, stimulating synthesis and secretion of testosterone. The cells in ovary respond to LH stimulation by secretion of testosterone, which is converted into estrogen by adjacent granulosa cells.

In females, ovulation of mature follicles on the ovary is induced by a large burst of LH secretion known as the preovulatory LH surge.

Residual cells within ovulated follicles proliferate to form corpora lutea, which secrete the steroid hormones progesterone and estradiol.

Progesterone is necessary for maintenance of pregnancy, and in most mammals, LH is required for continued development and function of corpora lutea. The name luteinizing hormone derives from this effect of inducing luteinization of ovarian follicles.

Follicle-Stimulating Hormone (FSH) as its name implies, stimulates the maturation of ovarian follicles. Administration of FSH to humans and animals induces "superovulation", or development of more than the usual number of mature follicles and hence, an increased number of mature gametes. FSH is also critical for sperm production. It supports the function of sertoli cells, which in turn support many aspects of sperm cell maturation. It causes ovulation and helps with the development of the breast and milk production.

Chapter 9

Female Reproductive Endocrinology

Copper (Cu) is one of the essential trace metals which are necessary in maintaining the functioning of living organisms. Studies have shown that complexes of copper (Cu^{2}+) with gonadotropin-releasing hormone (GnRH) are even more effective in the release of LH (luteinizing hormone a pituitary hormone that acts in the ovary to stimulate ripening of the follicle and formation of the corpus luteum) than native GnRH. Moreover, Cu-GnRH is more potent in inducing in vivo release of FSH (follicle stimulating hormone) than LH. Copper complexes with GnRH interact with GnRH receptors (GnRHR) and modulate intracellular signaling in the gonadotrope cells of the anterior pituitary. Copper also pays a significant role in maintaining normal fetus development.

A girl's first menstrual period (menarche) occurs several years after the beginning of the physical changes that define puberty. Puberty begins somewhere between the ages of nine and twelve for most girls. Menarche usually occurs between the ages of eleven and fourteen.

The external changes that are most obvious change in the outward appearance of a girl during puberty is the development of the rounded contours that distinguish the adult female profile from that of the male. This process begins with an increase in secretion of FSH by pituitary gland.

FSH stimulates the ovaries to produce estrogen (a collective noun referring to a group of chemically related female sex hormones produced by the ovaries). The estrogen, in turn, travels through the bloodstream to the breast, where it stimulates growth of the breast tissue. The pigmented area around the nipples (the areola) becomes elevated, and the breasts begin to swell as well as the result of development in the nipple area and an increase in fatty tissue, connective tissue, and blood vessels. Now keep in mind when we are dealing with pigment, we are dealing with copper; so we can see a high concentration of copper in the nipple area which is the main cause of the growth of the nipples. The milk-producing part of the breast (mammary gland) does not develop fully during puberty and indeed does not become fully mature and functional until pregnancy. The full development of the mammary gland doing pregnancy is a part of the secondary growth. This is why it is important for a female to have at least two children.

At the same time estrogen, which has a high concentration of copper, is inducing the development of fatty and supporting tissue in the breasts of the young girl (the pigment is stored in the fatty cells), a similar process is occurring in the hips and buttocks. These changes may be more pronounced in one region then in another in a particular woman. In some cultures the size and shape of the buttocks are important indices of femininity. In others greater emphasis is placed upon the breasts.

Another visible change at puberty is the appearance of pubic and axillary (underarm) hair. This hair is coarser and has more pigment. This is the same for the hair on the scalp, and its growth is stimulated partly by estrogen and partly by hormones secreted by the adrenal glands, also in response to a stimulating hormone (adrenocorticotrophic hormone, or ACTH) from the pituitary gland.

Estrogen also causes accelerated growth of the external genitals at puberty, including enlargement of the labia and associated structures. The clitoris enlarges under the stimulation of androgens from the adrenal glands. At the same time, there is a noticeable increase in the skeletal size, which is related to increased secretion of the size growth hormone by the pituitary gland.

Internal Changes. Estrogen specifically stimulates growth of the uterus and vagina during puberty. The muscular wall of the uterus enlarges, and its glandular lining also develops. The lining of the vagina is extremely sensitive to estrogen, and its thickness is proportional to the amount of this hormone present at any given time. Examination of a mucus smear containing cells from the vaginal walls is a simple and clinically useful test for determining how much estrogen is present. As the vaginal wall matures, the pH of the secretions that moisten its surface changes from alkaline to acid.

This may be the reason why young girls have a higher rate of getting pregnant since the vaginal fluid is highly alkaline giving the sperm a better chances of surviving. The acid kills the sperm, lessening its chance of survival. The pH scale is a commonly used index of acidity and alkalinity, based on hydrogen ion concentration. The scale runs from 0 to 14. The neutral point is 7. Values from 7 to 14 indicate an alkaline state, from 7 to 0 an acid state.) Under the influence of estrogen, the female pelvis enlarges during puberty and assumes different contours from those of the male pelvis.

Ultimately estrogen also prevent further growth of the long bones of the skeleton (counteracting the effects of the growth hormone), and usually no further increase in height occurs after about age seventeen in girls. An estrogen deficiency in late puberty may cause great height in a girl, for estrogen normally applies brakes to skeletal growth. Many other factors are, of course, involved in determining the height of a given individual.

Estrogen secretion not only increases in quantity during puberty, meditating the various external and internal physical changes that had just been described, it also takes on a cyclical pattern of secretion, which results in certain cyclical phenomena in the female, the most obvious of which is the menstrual cycle. The first menstruation occurs usually between the ages of eleven and fourteen years and may come as a shock to the girl who has not been prepared for it. For the first few years after the menarche the girl's menstrual periods tend to be irregular, and ovulation does not occur in each cycle. For some time after her first period, the girl is relatively infertile, but she can nevertheless become pregnant.

Use in female-FSH is primarily responsible for stimulating growth of the ovarian follicle, which includes the developing egg, the cells surrounding the egg that produce the hormones needed to support a pregnancy, and the fluid around the egg. As the follicle grows, an increasing amount of the hormone estrogen is produced by the cells in the follicle and released into the bloodstream.

Estrogen causes the endometrium (lining of the uterus) to thicken before ovulation occurs. The higher blood levels of estrogen will also tell the hypothalamus gland and the pituitary gland to slow the production and release of FSH.

Chapter 10

Male Reproductive Endocrinology

Copper (Cu) is one of the essential trace metals which are necessary in maintaining the functioning of living organisms. The current knowledge on the role of copper in animal reproduction is essential. The studies have shown that complexes of copper (Cu^{2}+) with gonadotropin-releasing hormone (GnRH) are even more effective in the release of (LH) luteinizing hormone, a glycoprotein hormone secreted by the adenohypophysis of vertebrates which stimulates hormone production by interstitial cells of gonads. Also known as interstitial-cell-stimulating hormone (ICSH). The native GnRH (the gonadotropin) is the male gonads or the male testicles. Moreover, copper-gonadotropin) release hormone is more potent in inducing in vivo release of FSH (follicle- stimulating hormone) than LH. Copper complexes with GnRH interact with GnRH receptors (GnRHR) and modulate intracellular signaling in the gonadotrope cells of the anterior pituitary.

Puberty begins somewhat later and lasts longer in boys than it does in girls. It is initiated at age ten or eleven in boys by the same pituitary hormones that mediate it in girls, FSH LH. But in males, LH is called "interstitial-cell-stimulating hormone" (ICSH) because of the difference in the site of its action.

ICSH reaches the interstitial cells of the testes (testicles) through the bloodstream and stimulates them, initiating the process of puberty. Little is seen in the way of striking external changes, but the interstitial cells begin producing the primary androgen (male sex hormone), a compound known as testosterone. Like estrogen, a female hormone, testosterone has a high concentration of copper as well. This one substance is essentially responsible for the development of all the physical changes (including development of the secondary sex characteristics) that occur during puberty.

External Changes. The first noticeable changes resulting from testosterone stimulation include enlargement of the testes and penis and the appearance of fine hair at the base of the penis. At first the penis increases in circumference more than in length. By age twelve it is still, on the average, about 1.5 inches long in the relaxed state and less than three inches long in erection.

As the testes enlarge, however, their capacity to produce testosterone increases, and at age thirteen or fourteen rapid growth of the penis, testes, and pubic hair begins. By this age, most girls will have had their first menstrual period, although they will not necessarily be ovulating yet.

Axillary hair does not appear until about age fifteen in most boys, and at the same time some fuzz appears on their upper lips. Adult beards do not appear for two or three more years, by age seventeen 50 percent of boys in the U.S. have not yet shaved, and many others shave only infrequently. Continued development of facial and chest hair, under androgen stimulation, continues beyond age twenty in many young men, along with recession of the hairline.

A very noticeable change that occurs during puberty, though strictly speaking an internal change, is the deepening of the voice, related to growth of the larynx ("voice box") in response to hormonal stimulation. The deepening of the voice may be gradual or vary abrupt, but on the average it occurs by age fourteen or fifteen. (Castration before puberty, of course, eliminates the source of testosterone that causes the voice to change. The famed castrati of Italy were men whose testes had been removed before puberty and who consequently kept their high-pitched voices all their lives. The voice of the castrated male and of the pubertal boy has always had its admirers, presumably because of a purer tonal quality).

Acne becomes a source of embarrassment to many boys at age fifteen or sixteen, as does another transient phenomenon, enlargement of the breasts, or gynecomastia. The latter occurs in about 80 percent of pubertal boys and is probably related to small amounts of female sex hormones produced by the testes.

Overall growth in both height and weight occurs during puberty. But whereas the girl develops fat deposits in the breasts and hips at this time, the boy's new weight is in the form of increased muscle mass, and he consequently develops a quite different physique. And, whereas the female pelvis undergoes enlargement at puberty, more striking expansion occurs in the boy's shoulders and ribs cage at this time. Of course, there is also a definite spurt in the linear growth rates of boys at puberty, and this process continues until age twenty or twenty-one, when the male hormones finally put the brakes on growth of the long bones of the skeleton. As previously noted, female sex hormones induce this "braking" process more rapidly, which accounts for the generally smaller height of woman. (This statement does not imply that a tall girl has a deficiency of female sex hormones. Other factors, including heredity, influence stature).

Internal Changes: Whereas the changes in external appearance just described give the appearance of biological maturity to the growing boy, significant internal changes must also occur before the capacity for reproductive activity is actually achieved.

The pituitary hormone FSH, which stimulates maturation of the ovum in the female, is also essential in the production of mature sperm in the male. With the increase in FSH secretion during puberty, germ cell in the lining of the seminiferous tubules of the testes begin to divide and differentiate into mature sperm. Mature sperm are first present in the ejaculatory fluid at age fifteen, on the average, but, as with the other changes at puberty, there is wide variation among individuals, encompassing the range from ages eleven to seventeen.

The organs that supply the fluids for the semen enlarge significantly during puberty. The prostate gland is particularly sensitive to the stimulation of testosterone. By age thirteen or fourteen the prostate gland is producing fluid that can be ejaculated during orgasm, but this fluid will probably not contain mature sperm. Just as the menarche does not necessarily indicate that a girl has ovulated, so the ability to ejaculate does not indicate that a boy is already fertile.

The testes serves two basic functions: One is to produce steroids hormones, and the other is to produce spermatogenesis, both regulated by the hypothalamus gland and the pituitary gland.

The neurons of the hypothalamus produce a protein hormone called gonadotropin releasing hormone (GN RH). GN RH is produced in a pulsatile fashion in the male. GN RN travels through the hypothalamo-hypohyseal portal system and acts on the anterior lobe of the pituitary gland simulating the pulsatile release of gonadotropins, luteinizing hormone (LH) and follicle-stimulating hormone (FSH). These gonadotropin hormones act on the gonads to stimulate testicular function.

Chapter 11

Hormone Glands and Growth

Copper is distributed through the glands in the form of hormones, whether we are tall or short, fat or thin, ugly or beautiful, brilliant or stupid; whether we have red hair or black hair or in certain cases, no hair at all; Whether our hair is straight or curly, our complexion peaches or mud, all depends a good deal upon the activity or the inactivity of the various ductless glands.

These glands are the fairy godmothers that give us freckles, big or little teeth, hands like hams, or several different kinds of fat distribution. So when there is fat in big blanket rolls around the lower part of the abdomen and hips that one little gland is asleep on the job.

Jewels: Our ancient ancestors called our glands, jewels or plumes. They were worn on crowns or hats to signify how important it was to have properly functioning glands. They understood the distribution of copper in the universe and how the universe was related to the proper functioning of the glands or jewels. They also knew that the jewels was responsible for the development and growth. The Sun distributes copper in the form of rays.

The thyroid: When it comes to the distribution of copper, the thyroid is one of the most important glands. It is because of its association with the gonads, or sex glands. It is the thyroid that has the greatest influence of all the glands on a healthy, happy life. For it is the direct index of one's available and useable energy. While of itself it does not actually produce energy, as do the adrenals, or the gonads, yet it accelerates, lubricates, and transforms our energies. It controls the speed with which chemical messengers, are carried by the blood to every cell in the body, where they govern the speed under which these cells function.

The thyroid needs iodine to function properly. The iodine helps to produce the T-4 to simulate the metabolism in the body tissue. It also produce T-3, which helps keep cholesterol in balance. The cholesterol is the most important hormone. It is where the pigment comes from.

The thyroid gland has much to do with the outer covering of the body. If you have a clear, beautiful, elastic skin; soft, lustrous hair, white pearly teeth; an active, questioning mind and ample energy, thank your thyroid gland.

In the sub-thyroid type, the eyes are often widely separated, sunken, dull and lacking in luster in distinct contrast to the bright, sparkling eyes of the thyroid-efficient type, or the over worked, startled, bulging eyes of the hyper active thyroid type as are noticed in many of our prominent screen artists.

Again, the person with a poor thyroid is usually short in stature thickset and small boned. Certain types of dwarfism are due to lack of thyroid secretion. Their hands are likely to be short and coarse, with thick, stubby fingers. The nails grow slowly, are usually brittle and sometimes cracked. Their teeth are improperly developed, enamel and dentine soft.

In conformity with a generally poor functioning condition, the hair of the poor thyroid type also suffers, lacking entirely the fineness and luster of the normal thyroid type. The hypo-thyroid parts with his or her hair easily. The eyebrows, also are inclined to be thin-in direct contra-distinction to the heavy, bushy eyebrows of the pituitary type. The skin is usually dry, harsh, and scaly like the hide of an alligator. It has not the genial warmth of the normal skin. The hands tend to get cold readily, and take on a bluish tinge closely resembling in this respect a prominent symptom of the sub-adrenal type, associated in low blood pressure.

Sub-thyroid people usually have an abnormally slow pulse rate and subnormal temperature sometimes as low as 96.6 or 97. They are easily tired, prone to contract any epidemic disorder that may be running around loose, and are chronic sufferers from, rheumatism, neuralgia, anemia, and shortness of breath. They also will have a low sex drive.

Women who lack thyroid efficiency usually suffer from delayed or scanty menstruation. Many women of normal thyroid power are definitely hypo-thyroid during pregnancy. They become dull, foolish, and slow during this period.

There is much more to be said about the thyroid gland, but by having your thyroid glands removed it will cause you to have hormonal changes that can cause one to become highly emotional, moody, shrewish, brain-stormy, quarrel some, fault-finding, clacking-tongued individuals of either sex, who" fly off the handle "with little provocation or for no provocation at all.

Hyperthyroid can occur from too much soda pop or acid foods which lead to low metabolic rate, weight gain, cold hands and feet, PMS, constipation, brittle nails, poor memory, headaches, migraines, and fibrocystic breasts. Under-active thyroid brings estrogen levels up and progesterone levels down, an under active thyroid may indicate a need for the minerals of Zinc and Iodine.

The Pituitary Gland. At the base of the brain, just below the roof of the nose, lies a tiny gland with a double lobe, known as the pituitary. This gland is placed by nature for its better protection in a little bony casket, inside the vault of the skull.

Secretion of the front portion of the lobe controls growth of the tendons and the bones, while the secretion of the back lobe regulates the tone of the nerves and involuntary muscles, and co-ordinates with the brain and sex organs in assisting their functioning powers. Among the wide range of functions of this gland are the maintenance of the splendid mental powers seen frequently among elderly people.

Those who lack pituitary secretion have many of the tendencies of the hypo-thyroid. They have little or no capacity for learning; they are mentally slow, apathetic or deficient; they are unable to maintain any protracted effort. They readily become discouraged and lose self-control.

In this type we find a peculiar obesity exhibited by so many women, with the girdle of fat around the lower abdomen and enormous fat distribution over the hips and buttocks, but with little or no excessive fat on the breast, neck or face. Among sub-pituitary boys there is a tendency toward the female contour, while the girls develop many masculine characteristics. Prior to puberty, there is also retarded growth of the bones, and small development of the sex organs and their function. They make very poor married companions, particularly if married to a highly sexed individual with normal libido. Indeed, complete impotence in the male and frigidity in the women are very frequently met in this type. In advanced cases we frequently find epilepsy, Mongolian insanity, and a tendency toward homo-sexuality.

When the study was done on those who had an excess of pituitary activity, it was an entirely different picture. Here we have great mental activity, with splendid initiative and fine brain power.

The men of this type are usually big boned and muscular, with large hands, feet and prominent knuckles. As a general rule, all people over six feet tall have a strong pituitary. This is the long, lean type, capable of eating the ordinary boarding housekeeper into bankruptcy, without having an ounce of additional flesh to show for it.

This type, as with a hyper-adrenal type, is apt to suffer from high blood pressure, as well as to be bothered by precocious sex urge childish libido.

The pituitary gland helps with the secretion of Human Growth Hormone (hGH) and Somatostatin (Growth Hormone Inhibitory Hormone). It mobilizes fat into the cells for fuel, helps sleep patterns, acts indirectly cartilage and bone growth, causes the adrenals to release cortisol and is influenced by stress (ACTH) Adrenocorticotropic hormone, a hormone normally produced by the pituitary gland, and the pituitary gland helps form the pigment melanin.

The Gonads or Sex Glands. It is now quite generally conceded that the internal secretions of the testes and the ovaries have more to do with shaping the character and personality of the individual than all the other glands put together. It is because the circulation of the blood ends up in the sex organs, according to our ancient ancestors and the Dogon Tribe of Mali, West Africa. This is the pathway of what our ancestors call the word. The clear blood or water blood (serum), which is the male element that passes through the right chamber of the heart, then passes through the lungs, liver and the spleen. Its number is 1 to 4 in red. The oily blood (plasma) which is a female element, passes through the left chamber of the heart then through the pancreas, kidneys, the intestines and sex organs, its number is 5 to 8 in white.

It can be said as a matter of truth, that the woman who has made history, has quite uniformly been well sexed. In point of fact, many of them were veritable nymphs, insatiable in their sex desires.

However, the average individual, with a normal love for companionship of the opposite sex, with a natural love for children, and with a well-developed sprit of camaraderie, is quite invariably well sexed.

The delightful charm of women, the soft, low voice, the rounded breasts and hips, the gentle maternal nature, exhibited for the whole world as well for her child, are reflexes of ovarian activity.

Absence of this sexual activity manifests itself usually in selfishness, and a certain tendency towards what almost might be described as sadism. It is this type that makes up the reformers, the prohibition enthusiasts, the antis against the activities of the multitude.

The Thymus Gland: This is the gland that make you meek, humble and causes you to have a high sex drive. This gland is situated in the chest, astride the windpipe, almost directly over the heart. It has been recognized as the gland of childhood. It will become calcified or diminish at the age of eight in people who have a low rate of copper or melanin. For people that lack copper, it will calcify early.

(This gland will not retard in people who have a high amount of copper in there system such as the indigenous people).

The Pineal Gland: The pineal gland is a cone shaped gland situated in the head, behind and slightly above the pituitary. It was originally thought to be the "seat of the soul." It helps with the process of Melatonin during the sleep state patterns, and helps with reproductive cycles. It needs sunlight to function properly, and it sends hormonal messages to the hypothalamus gland. Its greatest activity is manifested in childhood, it accelerates sex development. It is also a part of the third eye due to the fact that it is connected to the intuitive nature.

The Pancreas: The pancreas, while primarily a digestive gland, also secretes a ferment which has to do with the oxidation of sugar. When the island cells in the tail of the pancreas are affected by disease, or by toxemia, there develops an inability to oxidize sugar, which brings about the condition known as diabetes, from which thousands die early.

The Parathyroids: These four tiny glands are located in and around the thyroid gland. They control the metabolism of lime, and they influence the excitability of both nerves and muscles. The condition known as palsy is believed to be due to a defect in the functioning power of these glands.

Also, they are responsible for much restlessness and nervous excitability and the tremulous aspects in handwriting seen in many who show pre-senile symptoms. So if one has soft, fragile bones, extremely brittle nails, rapidly decaying teeth, and an early loss of hair, it is believed that quite frequently the Parathyroids are involved.

The Spleen: The spleen is one of the largest gland in the whole body, and is largely involved with the development of the red and white blood cells, as well as of the phagocytes that keep our system free from invading germs. Abnormal conditions of the spleen are mostly found among those who have suffered from attacks of malaria. In these conditions, anemia develops, together with lassitude, debility, loss of appetite, and general asthenia, or loss of strength.

Perhaps the most important gland out of all the glands we talked about for a woman is her sex gland, (her ovaries). That is the gland that give her beauty. It is the perfect functioning of these glands that gives to the woman her charm of femininity, her rounded contours and soft curves, her pitch of voice, the comparative absence of hair upon the face, neck, and breast.

The thyroid, as has been stated elsewhere, has a most important function to perform, in respect to maintaining the proper nutrition of the skin and in eradicating sources of poison that might constitute themselves dangerous or destructive to beauty, because the chief function of the thyroid is to secrete thyroxin, the purpose of which is to neutralize toxins in the blood stream. When this fails there is an accumulation of this poisonous debris in the system, the attempt to eradicate which through the skin produces pimples, boils, skin blotches, and eruptions.

The general hormones the internal secretion of the testicles and the ovaries, which is circulated through the system excite a very definite action in the organism. This action has to do not only with the mental, moral, and physical characteristics, but with growth as well.

Removal of the sex gland produces profound changes in the organism, as evidenced by alterations of bodily physique and temperament. When extirpation, or removal of the ovaries or testicles is done in the early life, the secondary sexual characteristics may fail to exhibit themselves in the normal manner, and thus infantile characteristics may be retained. Secondary sexual characteristics are associated with the influence of chemical substances produced by the ovary and testis respectively. Castration after puberty cannot modify profoundly the development of structures like the skeleton, which are already completed. Yet nothing more completely changes the entire human organism than the removal or atrophy of these, master glands of internal secretion.

The removal of the ovaries in the later stages of a woman's life will cause her to lose the ability to get moist or secrete vaginal fluid, therefore her ability to become aroused during sex can be lost.

Hypothalamus: (The Master Gland) the Hypothalamus is the primitive part of the brain. It controls the nervous system. It controls the blood pressure, hunger, gastrointestinal function, the feeling of fullness (after eating), thirst, water retention, bladder contraction, decreases heart rate, body temperature, sweating, sleep, wakefulness and alertness.

The hypothalamus initiates the anterior pituitary to secrete *TSH* (thyroid Stimulating Hormone).

Chapter 12

Copper in Food

Food is consumed to provide nutritional support for the body. It is usually of plant or animal origin, and contains nutrients, such as carbohydrates, fats, proteins, vitamins, or minerals. The substance is ingested to produce energy, maintain life, or stimulate growth.

The body doesn't need animal organs, carbohydrates, and proteins. (The body produces its own proteins). Vitamins are made of copper mixed with other mineral compounds. The fats are essential and are the most important source of food. It contains the pigment and is found in the skin of all plant life.

People think that the fats are bad for them, one reason they believe this is because of weight gain. If the body is not functioning right fats can be bad. The truth is fats carry the copper compounds known as vitamins, fats are what make up your cholesterol cells. When I speak of cholesterol I am speaking of (*sterols*) which is cholesterol from plant life. The brain is 75% cholesterol, so fats are a brain food. If the digestive system is properly working there would be no weight gain, but due to the toxic foods we are eating, we are clogging up our digestive system. The liver produces the green liquid bile that helps break down the fats, as well as the food in the digestive system, so the body weight can stay down. The problem is the liver gets clogged up and can't produce bile to keep the matter in the intestine flowing properly, so the body can be properly relieved.

The liver is the body's chemical powerhouse. It aids digestion, and keeps a store of glucose, vitamins, and minerals. It has many functions such as the removal of poisons from drugs and alcohol in the blood, but one of the most important function is exothermic, which means it give off heat to warm the body. The majority of copper that emerges from the foods is digested in the intestinal system, then into the blood, into the liver and less to the kidney and other tissues.

The liver is the central organ of copper homeostasis and is primarily responsible for the export of excess copper out of the body. The copper uptake into the liver does not appear to be highly regulated. In contrast, the export of copper from the liver is a regulated copper-dependent process.

Copper is an alkaline electrical source. This is important when it comes to the food we should be eating. All foods should be electrical and alkaline in nature, electrical and alkaline food is called life. You can't tell how much alkaline is in foods just by looking at it although, if it's green it is a good indication that it is alkaline. Dead foods are foods that are acidic which causes the body to decay. To get a better understanding, let's take a look at the molecular structure.

The Molecular Movement of Plants: The alkaline molecules move in the direction of the universe, during this movement the molecules in plant life will retain their green color. The temperature, water, sun light, and salt crystals also play a part in this state. The green pigment is called chlorophyll. Chlorophyll is the closest thing to the human blood. Chlorophyll is copper that has oxidized and developed salt crystals. This process is known as Verdigris. This is copper in its natural state of oxidization. When it comes to fruits and vegetables, this is also true. But when the molecules move in the opposite direction the plant life become more acidic. The fruit usually will fall off the stem or become soft. This is when fruit begins to rot *(enzymatic browning)* and the sugar content goes up. The colors change from green to whatever color they take on by nature, and the fruit becomes more pigmented. There is a point when the fruit will be at an alkaline state and will start to take on some sugar. This is also a good time to consume them. I am using fruit as an example because it is easier to determine the alkaline state or the acid content, although the color does not always determine the alkalinity. Another example is the leaves on trees changing their colors in autumn. In the summer time, the leaves will retain their green color as long as the temperature stays the same and the trees contain the salts crystals from the earth. When the temperature starts to change due to the angle of the sun, the trees will take on more water because they cannot extract the excess water due to less sun light. The molecules start to reverse in their movement of direction. The process causes the sugar content to go up and the leaves start to change colors. The color will be determined based on the amount of sugar the leaves consume.

The molecule movement takes place in all matter. It rotates in the direction of the universe at one period, then it will reverse its rotation at another period of time. Take a look at our own blood. Cold showers make the blood alkaline, while hot showers make the blood acid. When the blood is at an alkaline state the molecule movement is in the direction of the universe and just the reverse order when it's at an acid state. A balanced diet helps to maintain the pH balance of the blood. Results, however, take more than a couple of days to show up. If blood develops a more acidic condition, then the body will inevitably deposit the excess acid to another area of the body so that the blood will be able to maintain an alkaline condition. As this cycle continues, these areas will increase in acidity and some cells will die. The dead cells will then turn into acid. Acidic conditions inhibit nerve action whereas alkalinity stimulates nerve action. Hyperchloremic acidosis results from a loss of sodium bicarbonate. This base helps to keep the blood neutral. Both diarrhea and vomiting can cause this type of acidosis.

PH Scale: PH scale means potential of hydrogen or power of hydrogen, your blood (plasma) needs to maintain a pH of 7.35 to 7.45 for your cells to function properly. The reason is that all of the proteins that work in your body have to maintain a specific geometric shape in order to function, and the three-dimensional shapes of the proteins in your body are affected by the tiniest changes in the pH of your body fluids.

The pH scale ranges from 0 to 14. A liquid that has a pH of 7 is considered to be neutral. (Pure water is generally considered to have a neutral pH). The neutral state of copper is also green. Fluids that have a pH below 7 - like lemon juice and coffee - are considered to be acidic. And fluids that have a pH above 7 like human's blood and milk of magnesia (antacids) are considered to be alkali.

It's important to note that on the pH scale, each number represents a tenfold difference from adjacent numbers; in other words, a liquid that has a pH of 6 is ten times more acidic than a liquid that has a pH of 7, and a liquid with a pH of 5 is one hundred times more acidic than pure water. Most carbonated soft drinks (pop) have a pH of about 3, making them about ten thousand times more acidic than pure water. Please remember this the next time you think about drinking a can of pop. Pop is one of the main sources that cause the thyroid to go out of control especially in women. It can make the thyroid over active, causing weight gain.

When you ingest foods and liquids, the end products of digestion and assimilation of nutrients often results in an acid or alkaline- forming effect. The end products are sometimes called acid ash or alkaline ash.

Also, as your cells produce energy on a continual basis, a number of different acids are formed and released into your body fluids. These acids generated by your everyday metabolic activities are unavoidable; as long as your body has to generate energy to survive, it will produce a continuous supply of acids. There are two main forces at work on a daily basis that can disrupt the pH of your body fluids. These forces are the acid or alkaline-forming effects of foods and liquids that you ingest, and the acids that you generate through regular metabolic activities. Fortunately, your body has three major mechanisms (the kidneys, lungs and the liver) at work at all times to prevent these forces from shifting the pH of your blood outside of the 7.35 to7.45 range.

Kidneys and Lungs Regulate PH Level. The kidneys and lungs, play important roles in maintaining proper ph. The lungs excrete acids as carbon dioxide, and they do this without much effort or input from what we eat. The diet plays no direct role in the lungs' excretion of volatile acids. The kidneys' ability to excrete acids, however, is directly influenced by what we eat. On a balanced whole food diet, that includes ample amounts of fruits, vegetables, nuts, and seeds, the body is provided with the anionic (negatively-charged) alkaline mineral salts it needs for buffering excess acids (positively charged ions). Under these ideal circumstances, the kidneys are able to maintain the net acid-alkaline balance in proper proportion. An imbalanced diet high in animal protein, refined carbohydrates, caffeine, and processed foods, however, can force us into mild but chronic acidosis. As acidity rises, the kidneys must compensate by seeking and using the body's precious alkali reserves. As the alkali reserves become depleted, the body's systems become compromised and are forced into suboptimal functioning. Overall in our society, we consume a very imbalanced diet, high in acidifying foods. This imbalanced diet pushes us toward low-grade metabolic acidosis, to which the body's response is a withdrawal of calcium salts and other alkalizing mineral salts from the blood and tissues. The majority of fruits, vegetables, nuts, seeds, and spices have an alkalizing effect. Most grains are somewhat acid forming, as are many high protein foods. Refined sugar is acid forming, while natural salts is alkaline forming. Ideally, our diet should be composed of about 35% acid forming foods and 65% alkaline forming foods; however, the proportion of alkalizing foods should be higher when bringing the body back into balance, or when healing is required. For a comprehensive chart detailing the alkalizing and acidifying metabolic impacts of foods, you can find charts on the internet acid forming and alkaline forming food charts.

A related, but more well-known kidney condition that can result from all of this extra acid being neutralized by the body, are known as kidney stones, there two types; *uric-acid stones* and *calcium-oxalate stones*. (There are many types of kidney stones, formed by multiple causes). Now this is not intended to be an in-depth discussion on kidney stones, but if you have a history of stones, you should take note. Most kidney stones occur in people with inherited metabolic disorders, making them genetically predisposition to developing them. White men over the age of 40 are at even higher risk, especially if you drink a lot of coffee and soda, both of which are acidic, and both of which contain oxalate. Calcium binds to oxalate to form crystals in your urine. Calcium is the primary mineral used by the blood to neutralize acids such as uric, (protein metabolism), and acetoacetic acid, (fat metabolism and ketosis). By the way, if you're not getting enough calcium in your diet, it will be leached from your bones. The body will do whatever it takes to prevent blood-acidosis. If you stay well-hydrated these crystals in your urine will remain diluted and pass out of your system with no problem. But if you are chronically dehydrated, your urine can become highly acidic and these crystals will begin to combine into stones, especially if you have a family history. Once you develop one stone, your risk of developing more increases greatly.

You need to stay hydrated so your urine is always diluted. (This will help offset all the coffee or soda you may be drinking!). Even though one of the most common types of stones are composed of calcium-oxalate, most medical professionals agree that consuming dietary calcium will not increase your risk. Calcium is beneficial in so many other ways, for proper muscle function, neutralizing acids, and bone support. If you really want to take an extra step for your health, be sure to eat plenty of whole fruits and vegetables to help your body alkalinize itself. Even though citrus fruits are acidic, the body treats them as alkalinizing foods.

There are two types of acidosis, metabolic and respiratory. *Metabolic acidosis* occurs when your kidneys can't get rid of acid buildup or when your body gets rid of too much base. Bases neutralize acids, and vice versa. *Respiratory acidosis* occurs when your lungs do not properly eliminate the carbon dioxide (CO2). When CO2 builds up in your blood, it becomes more acidic. Metabolic acidosis starts in the kidneys instead of the lungs. It occurs when they can't eliminate enough acid or when they get rid of too much base. First and foremost, disease *thrives* in an acidic state and are less likely to live in an alkaline state. When we consume a predominantly acidic diet, we run the risk of acidosis. The body becomes too acidic and as a result, is forced to utilize its own alkaline mineral resources to maintain livable pH levels.

These minerals include sodium, potassium, magnesium and calcium. Your body is literally stripping itself of these minerals just to establish balance to neutralize the over-abundance of acid, and safely remove it from the body! As a result of acidosis, you may start to experience a build-up of acid in the cells, which then decreases energy production in those cells. Decreased energy production leads to an inability of the body to repair damaged cells and detoxify heavy metals, making you more susceptible to both chronic fatigue as well as an increased incidence of disease (cancer tumors thrive in an acidic environment).

Keep in mind that the pigment does not indicate the pH scale of fruits and vegetables although there are some fruits and vegetables that have a high alkaline state with a large amount of pigment, and there are some plant life that are high in acid with a high amount of pigment. The best time to consume fruits and vegetables is when they are in the green state, which is copper in its natural state of oxidation. (On the pH scale this state is neutral.)

The Flavor of life comes from Copper. Foods is where humans and animals have the pleasure of experiencing copper in the form of flavor and smell. The pigment is in the skin, the skin contains the fatty cells, the fatty cells produce the oils, and the oils contains the flavors and fragrance. This is one of the main reasons why fats are so important. Fats are also the key component in the photosynthesis process. It is the fats that captures the light from the sun. The fat also produces the melanin, and all of the other pigments in life.

Litmus paper. Litmus paper is one way of determining the acid or alkaline content in your food, you can buy it at some of your health food stores, janitorial supply stores, or online.

Chapter 13

What we Dispense is Copper

Copper is excreted in the lower hemisphere of the body, which is the intestines. The intestines is considered the lower brain. Its element is earth because it consumes matter or the foods we eat.

Matter produces the magnetic energy flow. This is where our emotions and feelings are derived.

One of the major organs in the digestive system is the liver. Its purpose is to produce copper in the form of bile. This bile is either green or yellow in color. Bile is an alkaline which is a source of copper. One of its functions is to digest fats, absorbs minerals like calcium, magnesium, and many more compounds. Without the absorption of the fats, we could not break down proteins. The fats break down the proteins. Once the liver produces the bile, the bile is then sent to the gallbladder. The gallbladder collects the bile and sends it to the small intestine where it is mixed in with the stomach acid to be neutralized and to protect the intestines from the acid. The bile is needed to break down the cells so the body can receive the nutrients we need to live.

In this section we are going to talk about the three main sources of copper that we dispense and that is urine, semen, and the feces.

Urine: Urine is a sterile liquid that is about 90% water. (Remember water is copper.) The rest is made up of a byproduct of what the body secrets through the kidneys, by a process called urination and is excreted through the urethra.

Color: Urine gets its yellow color from a pigment called urochrome. (Remember all pigment comes from copper.) That color normally varies from pale yellow to deep amber, depending on the concentration of the urine. Darker urine is usually a sign that you're not drinking enough fluid and the urine will become very strong and concentrated.

On the other hand if your urine is very pale, it means that you're either drinking a lot of water or, you're taking a diuretic (a drug that forces the body to get rid of excess water).

Urine can turn a rainbow of colors. Certain medications can turn the urine fluorescent green or blue, carrots can tint it orange, vitamins can give it a yellow hue, and an inherited disease called Porphyria can shade it the color of port wine.

Seeing a red color in the urine is a typical sign that there is blood in the urine. Blood in the urine can indicate something serious, such as an infection or cancer and it warrants a visit to your doctor or urologist.

The kidneys are the organs that regulates the PH level in the human body. It will either dispense acid or alkaline minerals in the form of urine. The PH level should be alkaline trace mineral and a water-soluble which are copper base.

Byproducts. Byproducts are very important, they consist of 2.5% of vital minerals, salts, hormones and enzymes. The other 2.5% is made of urea. Urea is sometimes used as a drug for treating both insulin dependence and non-insulin dependence diabetics.

There are a number of minerals in the byproducts of urine, here is a list of some.

Alanine: Is a colorless, crystalline, water- soluble amino acid, found in many proteins and produced synthetically. Alkaline Base.

Arginine: An essential amino acid that increases insulin secretion and is converted to urea in the liver by arginase. Alkaline Base.

Ascorbic Acid: A white, crystalline, water-soluble vitamin that occurs naturally in citrus fruits, green vegetables. Alkaline Base.

Allantoins: Is produced by oxidation of the uric acid and is the major excretory of purine degradation in many vertebrates. Medicinally it is used to heal skin ulcers and used in lotions, lipsticks etc. for its soothing effects. Alkaline Base.

Amino Acids: Is an organic compound that contains at least one amino group. It is the building blocks from which proteins are constructed.

Biotin: A crystalline, water-soluble vitamin of the B complex that is present in all living cells and functions as a growth factor and as a catalyst in carboxylation reactions. (Biotin is also called vitamin H), Alkaline Base.

Calcium: Calcium is lime which comes from the limestone. This is what our bones are made from. It also help regulate our nerve conduction due to its electrical principle as well as the heartbeat, muscle contraction, body heat, and many other physiological functions, Alkaline Base.

Creatinine: A crystalline product that occurs in the urine, muscle, and blood. Alkaline Base.

Cystine: Found in the bladder, it is an amino acid that occurs mostly in proteins, hair, wool, and the horns. The body produces this naturally. Alkaline Base.

Dopamine: Dopamine is formed in the liver. It has a high concentration of copper and the electrical principles are high. This is why dopamine is a catecholamine neurotransmitter in the central nervous system, retina, and sympathetic ganglia, acting within the brain to help regulate movement and emotions. Alkaline Base.

Epinephrine: Found in the kidney, epinephrine is a hormone secreted by the adrenal medulla upon stimulation by the central nervous system in response to stress, as anger or fear, and acting to increase heart rate, blood pressure, cardiac output, and carbohydrate metabolism.

Alkaline Base.

Folic Acids: Folic acid is a water-soluble vitamin that is converted to a coenzyme essential to purine and thymine biosynthesis. A deficiency causes a form of anemia. Alkaline Base.

Glucose: Glucose is sugar having several optically different forms, the common dextrorotatory from occurring in many fruits, animal tissues and fluids, ect...

Glutamic Acid: An amino acid that is obtained by hydrolysis from wheat gluten and sugar-beets residues, used commercially.

Glycine: Glycine is a water soluble, sweet crystalline. It is the simplest amino acid that is nonessential occurring in most proteins that acts as a neurotransmitter. Alkaline Base.

Inositol: A compound, derivative of cyclohexane, widely distributed in plants and seeds as phytin, and occurring in animal life, and present in the vitamin B complex. Alkaline Base.

Iron: Iron is a component of hemoglobin which allows red blood cells to carry oxygen and copper through the body. Alkaline Base.

Lysine: Lysine is a crystalline, basic, amino acid produced chiefly from many proteins by hydrolysis, essential in the nutrition of humans and animals.

Magnesium: An alkaline and an essential component of chlorophyll, which is the blood of plant life.

Manganese: Manganese helps to absorb and utilize other nutrients such as biotin, thiamin, ascorbic acids, and choline (a vitamin B-complex), carbohydrate and calcium. Alkaline Base.

Nitrogen: A component of all proteins, making it essential for life. It is also found in various minerals. Nitrogen is alkaline bases.

Ornithine: A formation of urea in the liver (bile). Alkaline Base.

Pantothenic acid: It is also found in plant and animal tissues, and rice bran. It is a part of the B complex of vitamins and is essential for cell growth. Alkaline Base.

Phosphorus Organic: Is essential for the creation of DNA, cell membrane, and for bone and teeth formation in the humans. So we can see it is a product of lime. Alkaline Base.

Potassium: Potassium is an alkali base. It is essential for growth of plant and animal life.

Proteins: Proteins are any group of complex organic macromolecules that contain carbon, hydrogen, oxygen, nitrogen, and usually sulfur and composed of chains of alpha-amino acids.

Riboflavin: Riboflavin is a vitamin B complex factor appearing as an orange-yellow, crystalline compound, derived from ribose and essential for growth. Alkaline Base.

Tryptophan: Tryptophan is an essential amino acid, crystalline, aromatic, released from proteins by tryptic digestion and a precursor of serotonin. Alkaline Base.

Urea: Occurring in urine and other body fluids as a product of protein metabolism. This is a water- soluble which is also used in medicine.

Vitamins: Vitamin B-6 is a water-soluble pyridine derivative, especially pyridoxine, which is a coenzymes in amino acids and synthesis and are important in protein and fat metabolism and in healthy nerve function. The vitamin is found in the liver, whole-grain foods, fish, yeast, and many vegetables. Alkaline Base. Vitamin B-12 is a deep red crystalline, water-soluble solid obtained from the liver, fish, eggs, oysters, and clams. It is good for the nervous system, Alkaline Base.

Zinc: Is essential to human and animal growth, Alkaline Base. Some of these vitamins and minerals are also used in medicines. As already stated in chapter 2, "What is Copper", copper is a water-soluble which is an alkaline base. Some of the minerals we dispense in our urine are either a water-soluble or an alkaline base, which is a byproduct of copper.

Urine Therapy. There have been a lot of research done on urine therapy. *Dr.L.Muldavis, in 1938, Royal Hospital* in London did some studies on urine in his book "*Your own perfect Medicine*". For almost a decade doctors and medical researchers have been proving in both laboratory and clinical testing that our own urine is an enormous source of vital nutrients, vitamins, hormones, enzymes and critical antibodies that cannot be duplicated or derived from any other source. They are using urine for healing cancer, heart disease, allergies, auto-immune diseases, diabetes, asthma, infertility, infections, and wounds. It is also used as a vaccine and has been shown to cure hepatitis, whooping-cough, hay fever, hives, migraines, and intestinal dysfunctions. Urine is used for many other medicines, and perfumes, the list goes on and on.

Embryonic Fluid: The embryonic fluid is made of urine. During the first stage of pregnancy the embryo is surrounded by water. After the first ten weeks, the fluid is replaced with the mother's urine. After twenty weeks the embryonic fluid is then replaced with the fetus urine, this fluid is called amniotic fluid, which is highly alkaline. The infant breaths in urine filled amniotic fluid continually, it is essential that fluid be breathed into the lungs in order for them to develop normally. Swallowed amniotic fluid also creates urine and contributes to the formation of meconium. Amniotic fluid protects the developing baby by cushioning against blows to the mother's abdomen, allowing for easier fetal movement and promoting muscular skeletal development.

Semen: Semen is a byproduct of bone marrow. The bone marrow is produced in the joints of the bones and is a byproduct of lime which is a trace element of copper. Remember the placenta is made from semen and the placenta is copper.

Feces (Fecal Matter): Copper is the product of many of the foods we eat especially in the fruits, vegetables, grains, beans, and herbs so it would be natural for us to dispense copper.

Chapter 14

Copper is Disease Free

The metal Copper, is known to be disease free. Copper is an alkaline and all of your alkaline base compounds are a form of copper. It has been known for some time that copper can be used to kill germs and bacteria.

The metal copper, has been mixed with other metals, such as tin, to create brass for the use of door knobs, face plates, plumbing pipes, tables and even utensils, etc. There is a lot of research that is taking place right now dealing with the study of copper being a disease free metal in hospitals and in other social public places.

With that being said, we have to also look at the fact that we, as a people, are copper beings and should also be disease free. So why do we as copper beings, get diseases? I would say that one of the main reasons is, we are consuming food that has a small amount of copper. Since copper is an alkaline base compound, if we stick to alkaline foods we should be okay.

So what foods causes disease? Any foods that are acid base cause diseases. Acid forms mucus which is the causes of all diseases. Mucus builds up in the body and becomes calcified or harden causing blockage.

Once that blockage takes place, it stops that part of the body from getting all of the nourishment and electricity it needs to stay alive.

Acid is the opposite of alkaline. Although we need some acid in our system to form the white blood cells, it is when we have too much acid in our system that it becomes dis-ease. Copper is an Alkalis and Bases, they can dissolve in water. These ions react with the hydrogen ions (H+) in alkaline to cancel out acidity.

The first recorded medical use of copper is found in the "*Smith Papyrus*", one of the oldest books known. The papyrus is an Egyptian medical text, written between 2200 and 2600 B.C., which records the use of copper to sterilize chest wounds and to sterilize drinking water. Other early reports of copper's medicinal uses are found in the "*Ebers Papyrus*", written around 1600 B.C. The "*Ebers Papyrus*" documents pertained to medicine practiced in ancient Egypt and other cultures that flourished many centuries earlier. Copper compounds were recommended for headaches, "trembling of the limbs", burn wounds, itching and certain growths in the neck, some of which were probably boils.

Forms of copper used for the treatment of disease ranged from metallic copper splinters and shavings to various naturally occurring copper salts and oxides (which turns copper green). A "green pigment" which is spoken of was probably the mineral, malachite, a form of copper carbonate. It could also have been chrysocolla, a copper silicate, or even copper chloride, which forms on copper exposed to seawater. In the first century A.D., Dioscorides, in his book, *"De Medica"*, described a method of making another green pigment, known as verdigris, by exposing metallic copper to the vapors of boiling vinegar. In this process, blue-green copper acetate forms on the copper surface. Verdigris and blue vitriol (copper sulfate) were used, among other things, in remedies for eye ailments such as bloodshot eyes, inflamed or "bleary" eyes, "fat in the eyes" (trachoma), and cataracts.

Copper is recommended for the treatment of leg ulcers associated with varicose veins. To prevent infection of fresh wounds, the Greeks sprinkled a dry powder composed of copper oxide and copper sulfate on the wound. Another antiseptic wound treatment at the time was a boiled mixture of honey and red copper oxide. The Greeks had easy access to copper since the metal was readily available on the island of Kypros (Cyprus) from which the Latin name for copper, cuprum, is derived. In the new world, the Aztecs also used cooper for medical purposes. Don Francisco de Mendoza commissioned two learned Aztec Indian physicians to record the pharmacological treatments known by the Aztecs at the time of the conquest. For the treatment of "Faucium Calor" (literally, heat of the throat, or sore throat), they prescribed gargling with a mixture of ingredients containing copper.

Turning to more modern times, the first observation of copper's role in the immune system was published in 1867 when it was reported that, during the cholera epidemics in Paris of 1832, 1849 and 1852, copper workers were immune to the disease. More recently, copper's role in the immune system has been supported by observations that individuals suffering from Menke's disease (an inherited disease in which there is defective copper absorption and metabolism) generally die of immune system- related phenomena and other infections. Further, animals deficient in copper have been shown to have increased susceptibility to bacterial pathogens such as Salmonella and Listeria.

Evidence such as this led researchers to suggest strongly that copper compounds not only cure disease but also aid in the prevention of disease.

Since 1934, it has been known that individuals suffering from such disease as scarlet fever, diphtheria, tuberculosis, arthritis, malignant tumors and lymphogranulomas exhibit a lack of elevation of copper in their blood plasma. Since then, the list of maladies bringing about such elevation has been extended to fever, wounds, ulcers, pain, seizures, cancers, carcinogenesis, diabetes, cerebrovascular and cardiovascular diseases, and irradiation and tissue stresses, including restricted blood flow.

This suggests that this redistribution of copper in the body has a general role in responding to physiological diseases, or injury stress. On the other hand, the elevation of copper in the affected organ has led some to postulate that it was this excess of copper that caused the disease. Nonetheless, this elevation of copper in a diseased state, is suggested to account for the natural synthesis of copper- dependent regulatory proteins and enzymes in the body required for biochemical responses to stress. It may be that these natural copper complexes expedite the relief of stress and the repair of tissues. Thus, it appears that in addition to the anti- bacterial and anti-fungal activity of inorganic copper compounds as recognized by the ancients, metallo-organic complexes of copper have medical capabilities that are fundamental to the healing process itself.

Copper is known to be an essential element in human metabolism. However, copper does not exist in the body in measurable amounts in ionic form. All measurable amounts of copper in the body exist in tissues as complexes with the organic compounds of proteins and enzymes. Therefore, it has been concluded that copper becomes and remains intimately involved in body process. Some copper complexes serve to store copper, other to transport it, and yet others play important roles in key cellular and metabolic processes. Studies into the roles that these copper complexes play and mechanisms of these roles have further confirmed that copper enters into the prevention and control of a number of disease states in the body.

Ulcer and wound-healing Activities of Copper Complexes. It has been demonstrated that copper complexes such as copper tryptophan, markedly increase healing rate of ulcers and wounds. For example, copper complexes heal gastric ulcers five days sooner than other reagents. Further, it has been shown that, whereas non-steroidal anti- inflammatory drugs, such as ibuprofen and enefenamic acid suppress wound healing, copper complexes of these drugs promote normal wound healing while at the same time retaining anti-inflammatory activity.

Anticonvulsant Activities of Copper Complexes. The brain contains more copper than any other organ of the body except the liver. This fact suggests that copper plays a role in brain functions. With reports of seizures in animals and humans following the protected consumption of copper-deficient diets, it was reasoned that copper has a role to play in the prevention of seizures. It was subsequently discovered that organic compounds that are not themselves anti- convulsants exhibit anticonvulsant activity when complexed with copper. Further, it was found that copper complexes of all anti-epileptic drugs are more effective and less toxic than their parent drugs.

Anticancer Activities of Copper Complexes. As early as 1912, patients in Germany were treated for facial epithelioma with a mixture of copper chloride and lecithin. Success of such treatment suggested that copper compounds have anticancer activity. Work at the University of Liverpool in 1913 demonstrated that subcutaneous and intravenous injections of a copper salt or colloidal copper softened and degenerated carcinomas transplanted into mice. In 1930, work in France indicated that injections of colloidal copper mobilized and expelled tumor tissue. Recent work with mice in the USA has shown that, indeed, treatment of solid tumors with non-toxic doses of various organic complexes of copper markedly decreased tumor growth and metastasis and thus increased survival rate. These copper complexes did not kill cancer cells but caused them to revert to normal cells.

Anticarcinogenic Activity of Copper Complexes. Based on work in the treatment of cancers using copper complexes, researchers have found that these same complexes may prevent or retard the development of cancers in mice under conditions where cancers are expected to be induced.

Radiation Protection and Radiation Recovery of Copper Complexes. Ionizing radiation, such as that used in the treatment of cancer, has been shown to induce massive systemic inflammation.

Ideally, such radiation-induced injury might be prevented or ameliorated by chemical repair mechanisms in the body. Thus, pharmacological approaches to the repair of radiation-damaged tissue are needed. As early as 1984, copper metallo-organic complexes have been shown to have radiation protection and radiation recovery activities. They are capable of causing rapid recovery of immunocompetence and recovery from radiation induced tissue changes. The mechanism of this activity appears to be tied to the ability of certain copper complexes to deactivate the superoxide, or "free," radicals liberated by ionizing radiation. In addition, since radiation has the capability of breaking the bonds of natural copper enzymes in the body, supplementing these complexes may also have Anti-carcinogenic activity. It is suggested that there would be merit in using these complexes because they may also have Anti-carcinogenic activity. It is suggested that there would be merit in using copper complexes. In the treatment of cancer and in particular, treating patients undergoing ionizing radiation therapy for cancer, accidental exposure to radiation, and astronauts undertaking space travel.

Heart Disease and Copper Complexes. Numerous studies have drawn attention to the relationship between copper deficiency and heart disease. First observed in rats in 1936, this effect has now been traced to both a deficiency in copper and an imbalance in the copper-to-zinc ratio in the body. There has been work done on this subject by Dr. L.M. Klevay at the U.S. Department of Agriculture, Human Nutrition Research Center in 1973. This research has led to the postulation that copper has a direct effect on the control of cholesterol. In continuing work published in 1975, he theorized that a metabolic imbalance between zinc and copper-with more emphasis on copper deficiency then zinc excess is a major contributing factor to the etiology of coronary heart disease.

Subsequent work by other investigators has shown that copper complexes also can have a valuable role in the minimization of damage to the aorta and the heart muscle as oxygenated blood reperfuses into tissues following myocardial infarction. This action is based on the anti- inflammatory action of copper complexes. These and other studies suggest the use of copper dietary supplements as a means of preventing and controlling such diseases as atherosclerosis (a form of arteriosclerosis), coronary heart disease, aortic aneurysms and myocardial infarction can be treated with the proper amount of copper intake. It has been speculated that the reason that the heart attack rate in France is lower than in America is because France prepare their white wine with the skin of the grape intact; the pigment in the skin has a high rate of copper.

Copper test done on Copper Surfaces. Influenza A is a virus, which easily travels through the air after respiratory discharges (sneezing, or coughing). Airborne virus particles quickly settle on any surfaces in the vicinity. Active virus particles can then be distributed to additional parts of a facility by touch, potentially putting more people at risk. 2,000,000 virus particles of influenza A were placed on a stainless steel surface and after six hours, 1,000,000 particles were still active. After24 hours on the stainless steel, 500,000 particles were still active. When the same amount of influenza particles was placed on a copper surface, only 500 remained after six hours. No active flu particles could be detected on the copper surface after 24 hours.

Copper test done on Escherichia Coli. Escherichia Coli is a serious pathogen causing hemorrhagic colitis. It has been responsible for several large-scale outbreaks in recent years. E coli is able to survive in a range of environments, under various conditions. The risk of infection from contamination surfaces is recognized, especially due to the low infectious dose required. In this study, a high concentration (10x7 cells) of E coli was placed onto different metals and survival time measured. Results showed E coli to survive for over 28 days at both refrigeration and room temperatures on stainless steel. Copper, in contrast, has strong antibacterial properties (No bacteria can be recovered after only 90 min exposure at 20 degrees Celsius.) But its poor corrosion resistance and durability make it unsuitable for use as a surface material. Other copper-containing alloys, such as copper nickels and copper silvers, have improved durability and anticorrosion properties and greatly reduce bacterial survival times at these two temperatures after 120 min at 20 degrees Celsius and 360 min at 4 degrees Celsius, no E coli could be detected on a copper and nickel compound with a 73% copper content. Use of a surface material with antibacterial properties could aid in preventing cross-contamination events in foods processing and domestic environments, if standard hygiene measures fail.

Ultraviolet Light Kills Germs. It should not be a surprise that ultraviolet light kills germs since the light is copper. Remember the sun produces the ultraviolet light in the form of rays. The rays produce the seven different colors that we cannot see which come together, and when they come together, they produce the white light we can see. This is why we must keep in mind that copper is the source of all colors, pigment, water (which carries the pigment copper in it), and light as well.

The ultraviolet light kills about 99% of the germs. In fact, they are using ultraviolet lamps in sewage treatment plants, hospitals to clean rooms and in many other areas where it is crucial because a large amounts of bacteria, viruses and spores exist.

Trace minerals: Trace minerals are minerals that are copper bases and are used to fight cancer.

The fact that we are copper beings, we were created to be disease free!

Chapter 15

Copper and the Photosynthesis Process

Photo means light or that which is produced by light. The word synthesis means the putting together. So photosynthesis is that which is produced by light coming together. But what is light coming to together to do? To convert light energy which is captured from the sun by all life. That light energy is changed into chemical energy that can be used as fuel for the organism in the human body, as well as the animal and plant life. This chemical energy has an effect on the mind, as well. This light is color, which also vibrates in the form of sound, and this sound vibration has an effect on all life.

The light energy is copper. The sun is made of copper because the sun's ultraviolet rays are colors. These colors we can sometimes see in water vapor in the atmosphere when they break apart in wavelengths creating a rainbow. When the wavelength come together they form the white light we see. Life absorbs this light energy and changes it into chemical energy. The human mind is made of electrical chemical impulses. This is true for the universal mind as well and copper is the source. Before we talk about the photosynthesis process, let us talk about the colors and how they vibrate.

Color Vibrations: The varying of light contains all the splendors of the universe. As the velocities decrease from white light, which travels at the speed of 186,000 miles a second through violet, indigo, blue, green, yellow, orange, and red to black, which goes down to the speed of 140,000 miles per second. It is by the varying movements of these velocities that the eye is affected by sensation known as color.

The molecular constitution of a body determines the character and the speed of the light vibrations it returns to the eye, and thus give each body its own characteristic color. Hence, the term "color" is used to denote the different appearances that matter presents to the eye independent of its form.

Black is composed of equal parts of red, yellow, and blue. White is composed of five parts of red, three parts of yellow, and eight parts of blue. Normal or natural gray is composed of white and black in equal proportions. Black, therefore, means the result of all colors, while white signifies the reflection of all colors; and each color in its turn is but a mode of motion, or the varying sensation that we experience when these vibrations impinge upon the optic nerve.

Each color is, therefore one of the manifestations of vibrations (which produce their own sounds) and all vibrations manifests in corresponding color, the color being merely an indicator of occult chemical activity.

The color indicates not only the quality but the value of any chemical substance by reason of its own particular rate of vibration, and all objects of any particular color have that vibratory activity and value which pertain to that color. As the color is, so is the vibration, as well as the sound that it produces.

Nature's scale of vibration is very wide in its extent. It commences with sound, then merges into thermal heat waves, and these vibrations climb the vibratory scale as the temperature increases and merge into the vibrations of the radiant heat wave in the infrared which reach up to the visible red of the light spectrum.

The sensation of color, therefore relies on the number of vibrations of light ether, just as the pitch of a note depends on the number of vibrations of the sounding body. The number of vibrations for each color is constant. These seven different rates of vibration that we know as colors make up the visible light spectrum. Beyond these are series of vibrations known as the ultra-violet. These extend up to those vibrations designated as radioactivity, of which radium and the x-rays are best known examples. This ends both the light spectrum and nature's vibratory scales so far as present knowledge goes. The quality and value of all forms of energy are due to their rates of vibration, it follows the color that will indicate the quality of the vibration energy.

Each color has a distinct frequency or vibration due to wavelengths. Red has a larger wave length then blue, which is proportionately much shorter. That is why the quick notes of the drums arouse the emotions of the savage beast. They express in sound what red expresses in color, while the neutral notes of the flute or horn have a quieting influence. They express in sound what blue expresses in color.

These manifestations are recognized as they are perceived by the different nerves, for the mind of man translates the impressions of the world into facts of consciousness and thought by means of the nerves. All these varying rates of vibration differ only in direction, rate, and frequency, and are interpreted according to the different nerves or groups of nerves physically attuned to them or organized to select and respond to especial manifestation of vibratory activity.

Color is both physical and psychological in its effect upon the mind. The physical effect is a chemical one; the psychological effect is psychic which is electrical. The nervous system reflects its disturbances upon the mind (This is the sixth sense.) Hence the sensation of pain or pleasure and the emotional states that accompany them. This is true of all the colors. Primary colors are radical, elemental, and fixed in their vibrations or wave lengths, and hence, when once the effects of the sensations which they produce on the mind are known, their uniformity can always be depended upon.

Red is thermal and a stimulant and blue is electrical and depressing. They act uniformly on all forms of life. The spectrum analysis proves that the seven colors of light are made up of vibrations or waves lengths of mathematical exactness.

As all vibratory activity expresses itself in form, color, sound, taste, and smell covering the senses, it follows the energy and is always of that particular color, shade, or tint that belongs to that particular rate of vibration. This is true of those colors, shades, and tints that are perceptible to visible light, but also of the finer forces of nature that transcend the physical senses.

All energy of whatever quality is continually vibrating and, by reason by this vibration, it assumes the color, shade, or tint belonging to that specific rate of vibration.

The primary colors are red, blue, and yellow. All the other colors are combinations of these colors in certain proportions, red is the physical color and is the color of all physical energy. The various shades of red have their various significations, but, generally speaking, red is the physical color.

Red is also the color of the will. Wherever the will is expressed, it assumes the red color because will is the out-flowering or manifesting principle, the extension or expression of the self into manifestation. Therefore, it must express itself in and through the physical; consequently, it becomes red in color.

The second primary color is blue, the color of emotion. All energy vibrating on the higher octave is either blue or red, according as it is positive or negative, blue being the negative, feminine, or magnetic side of matter, and red the positive, masculine, or radiant side. The bluer the matter may be, the more magnetic it is, while the redder it may be the more electrical it is.

The third primary color is yellow and it is the color of the mental plane. Everything vibrating on the mental octave is yellow. The darker the yellow, the more gross and material is the character of thought; the lighter it becomes, the more spiritual it is. The clearer the yellow is, the more purity of thought. The brighter the yellow, the more brilliancy of the mind is indicated.

Yellow or orange symbolizes the highest function and power of our nature and preeminently typify the "scientific" temperament.

White is really the combination of all the seven colors. They are all found to unite in forming the white ray. The prism has the power of breaking up the white light into seven prismatic rays, therefore white light is really the combination of those rays.

Black is the source of dark energy matter. It is where the white light is emanated from. It means peace, harmony in its dark pure state and activity in its light white state.

Scarlet is the color of anger, the color which the astral body assumes when it is in an intensely angry condition. Anger is really the forcible action of the will moving outward in a very positive manner. This is what produces the state of anger.

The color of ether is pink, although its vibration is so intense that very few are able to see the vibration; but ether is always pink.

Crimson is the color of affection and human feeling. It is the self-relative color, because our affection is given to persons on account of their particular relation to us. Rose is the color of life, and as we approach the physical it is redder. As life is brought under the influence of the emotions, it takes on whatever color is found, blended with this red or rose color.

Indigo blue is the color of occultism. The dark indigo partakes of the element of screen, while the pure indigo is emblematic of pure occultism.

Violet is the color of magic because it is so far above the ordinary rate of vibration that it has the power of neutralizing and even transforming those rates into its own, thus giving the power of alchemy.

Purple being blue mixed with red, the positive aspect of emotion, is the color of mastership, and indicates the master.

Lavender, which is a great deal of white mixed with purple, indicates the master on the astral plane, but merging toward the spiritual.

Green is the color of action. It is expressed by minor chords, and is positive expression of the inner being as expressed through action. Unselfish action, action which is purely altruistic in its character, which has no relation whatever to the individual is a clear emerald green, and the nearer this approaches to the spiritual the more lighter it becomes.

Brown is, in a sense, a mystical color, in that it indicates the presence of white, red, and black in certain combinations. As the shade is, so will its influence be.

As all the colors are included in the white, so is all matter built up of simpler or elemental substances and all changes are due to distribution and aggregation of the elemental matter.

Motion and the magnetic field is the primary cause that gives rise to color and all other phenomena in existence.

In order to trace the origin and progress of any effect, or visible or invisible, the aim must be to determine the movement or movements which brought the phenomena into existence.

The principle of vibration permeates the whole science of radiation or motion and may be stated in a general way by saying that a body absorbs waves that are of the same period as those which it emits. It is itself in vibration.

The essential condition is, therefore, that the receiver shall be in the same key or wave length as the sender or origin of movement.

The original wave gives energy to the wave upon which it impinges. While one is gaining energy, the other is losing it and continues until the process of absorption is completed.

The principle applies to light, heat, sound, color, or energy of any kind where it is distributed by a process of radiation.

All frequencies that come from sounds, colors, smells, taste, and motion, are vibrations that move on the electromagnetic spectrum.

***How the color of light affects Photosynthesis*.**

Light: Light is essentially a wave on the electromagnetic spectrum. A wavelength measures the distance between each wave. Visible light, for instance, has a wavelength of 380 to 750 nanometers (a billionth of a meter.) Purple is on the shorter end, and red is on the longer end, and on either side of them are ultraviolet and infrared. Shorter wavelengths have higher energy. White light is composed of a combination of the visible spectrum, but different wavelengths may not cover all areas equally.

Pigments. Life and organisms can only absorb certain wavelengths of light, usually all of the other colors but not their own. The visible light that is reflected back is their own color, which gives photosynthetic organisms their color. Chlorophyll A, which is green in color for instance, absorbs red and blue-violet but reflects mostly green, giving plants with chlorophyll an intense green color. Chlorophyll B varies slightly, reflecting green and yellow light. The reason some deciduous trees and shrubs begin to change color in the autumn is because they stop making food during that time, and the chlorophyll slowly disappears.

Function. Plants, algae, sea weed (green in color) and cyanobacteria (blue-green in color) are dependent upon the absorption of light so that they can drive cellular metabolism or release hormones at specific times. They are adapted to use a specific pigment for their unique situations and environments. Algae in shallow water, for instance, may use Phycobilin (any class of red blue- green pigment found in the red algae cyanobacteria) to easily capture red-yellow light, but algae deep in water tend to capture blue-green light where it is most prevalent. However, colors in the green wavelength range only need to be absorbed at low levels or not at all, so green is the most abundant color reflected by photosynthetic organisms, which gives the earth a familiar green color. Since blue has a shorter wavelength and provides more energy, it is necessary energy. An organism may grow or work at different variables of efficiency depending on the wavelength of light, but they also can have many different kinds of pigments within each light harvesting complex, preventing stunted or lethargic functionality.

We can even see in all organic life that copper is the source for proper growth.

How does photosynthesis work: Plants are the only photosynthetic organisms to have leaves (and not all plants have leaves) A leaf may be viewed as a solar collector crammed full of photosynthetic cells. The raw materials of photosynthesis, is water, and copper enters the leaves by absorbing the ultraviolet rays from the sun this causes the plant to become hot and dry seeking water which is extracted from the earth through the roots of the plant. The water is then transported to the leaves through specialized plant cells known as xylem.

The human skin is also a solar collector crammed with melanocytes cells. These cells help produce the pigment call melanin that filters out some of the ultraviolet light that is radiated from the sun. The rest of the light is absorbed to produce the light energy which is made from copper and other compounds which make up the molecules the body needs for development and growth. (The light rays are rays of color).

What is really being absorbed in the leaf and the human skin is copper. Remember the ultraviolet light is color rays and all colors comes from copper.

Land plants must guard against drying out (desiccation) and so have evolved specialized structures known as stomata to allow gas to enter or leave the leaf. Carbon dioxide cannot pass through the protective waxy layer covering the leaf (cuticle), but it can enter the leaf through an opening (the stoma which means hole) flanked by two guard cells. Likewise, oxygen produced during photosynthesis can only pass out of the leaf through the opened stomata. Unfortunately for the plant, while these gases are moving between the inside and outside of the leaf, a great deal of water is also lost.

7-Dehydrocholesterol: In the human skin, 7- Dehydrocholesterol (cholesterol which is your fatty cells) is a zoosterol that functions in the serum as a cholesterol precursor, and is converted to vitamin D-3 in the skin. *(This process is really photosynthesis. There is no such thing as vitamin- D).* The sun rays is absorbed through the skin heating up the fatty cells and causing the fatty cells to loosen and move carrying the light rays, which is converted into energy, to all parts of the body. The presence of this compound in the human skin enables humans to manufacture vitamin D-3 from the ultra-violet rays in the sun light, via an intermediate isomer pre-vitamin. (Vitamins are copper mixed with other mineral compounds). The skin contains two primary layers, the inner layer called the dermis, composed largely of connective tissue, and the outer thinner epidermis. The highest concentrations of 7-dehydrocholesterol are found in the epidermal layer of the skin. The production of pre-vitamin D-3 is, therefore, greatest in these two layers, whereas production in the other layers is reduced.

Lipids: Lipids are very diverse in both their respective structures and functions. These diverse compounds that make up the lipids family are so grouped because they are insoluble in water. They are, however, soluble in other organic solvents such as ether, acetone and other lipids. Major lipid groups include fats, phospholipids, steroids and waxes. The photosynthesis process that take place in the fatty cells is sometimes called lipids.

Lipids fats are composed of three fatty acids and glycerol, an odorless, colorless syrupy liquid. These triglycerides can be solid or liquid at room temperature. Those that are solid are classified as fats, while those that are liquid are known as oils. Fatty acids consist of a long chain of carbon. This carbon is really copper.

Lipids absorbs the ultra-violet waves from the sun. The fatty cells take in the color rays which are mixed with other compounds in the fatty cells that produce the vitamins and hormones that the body need to build itself. The heat from the sun causes the fatty cells to break up and lose their ions, causing them to become charged which causes the movement of the cells. The cells move to all part of the body.

Lipids and Steroids: Steroids have a copper backbone that consists of four fused ring like structures. Steroids include cholesterol (known as 7 Dehydrocholesterol), sex hormones (progesterone, estrogen and testosterone) and cortisone. The steroids are produced through the process of photosynthesis.

Melanocyte: Melanocyte is a type of cell prominent in the dermis (middle layer of the skin) that produces melanin, the pigment that gives color to the human skin as well as protects the skin from ultraviolet light damage. There are two types of melanin: the dark brown pigment eumelanins and the red/yellow pigment pheomelanin. The skin contains the same number of melanocytes no matter what the individual's natural skin color. The melanocytes in darker skin are more active than those in lighter skin. The eyes and hair also contain melanocytes.

Melanogeneisis: The exclusive role of melanocytes is to produce melanin (melanogeneisis), a somewhat sequential process. To prepare for melanogeneisis, the body produces the enzyme tyrosinase. Genetic encoding regulates this process. Tyrosinase initiates conversion of the amino acid tyrosine, which synthesize from dietary cholesterol from plant life (sterols) and which the melanocytes store, into dopaquinone. The dopaquinone forms the pigments eumelanins and pheomelanin, which collectively comprise melanin.

Ultraviolet light Exposure. Exposure to Ultraviolet light, notably sunlight, initiates a sequence of hormonal and chemical events that stimulate melanocytes to produce melanin (Melanogeneisis). Sunlight (or other ultraviolet light exposure) sometimes can damage the cells of the skin, especially for those who have little amounts of pigment or melanin. The damage activates causes the natural repair mechanisms within the cells, which releases chemicals into the bloodstream that travel to the pituitary gland. In response the pituitary gland releases melanocyte-stimulating hormone (MSH), to build melanocytes. Melanocytes pass packets of melanin molecules to the keratocytes, which carry them to the outer layer of the epidermis as they migrate upward.

The resulting skin color depends on the mix of eumelanin and pheomelanin. The melanin in the light skin contains more pheomelanin than eumelanins. In darker skin, the balance tips the other way with the melanin in dark skin containing more eumelanin than pheomelanin. In the epidermis, melanin protects the skin from damage by absorbing ultraviolet light. The darker the skin, the less ultraviolet light penetrates the epidermis. In general, it takes about a week of regular sun exposure to generate a tan adequate to begin protecting the skin from further sun damage, though the tan itself signals sun damage.

Sun Light Exposure to the Eyes. The process of Photosynthesis is to consume the energy from the greatest power house in the universe, the Sun (light) and the most effective consumption of this light is through the eyes known as sun gazing. This is what will activate the melanopsin, is a photopigment found in specialized photosensitive ganglion cells of the retina, that are involved in the regulation of circadian rhythms, pupillary light reflex, and other non-visual responses to light. In structure, melanopsin is an opsin, a retinylidene protein variety of G-protein-coupled receptor.

Melanopsin is most sensitive to blue light. Modern day scientist tells us that looking into the sun will cause you to go blind. This is a myth. The truth is our ancient ancestors have been sun gazing for thousands of years. Looking into the sun to consume energy should be achieved slowly. In the beginning stages of sun gazing, you should look into the sun for only a few seconds at a time and it should be done during the first two hours in the morning, at sunrise, or the last two hours, during sunset. You can increase the exposure over time, and you will start to notice an abundance of energy in the midbrain region (known as the third eye) which will increase your concentration, sight, and awareness.

Vegetation is a byproduct of the sunlight and water. The sun is made from copper and water. The plant life consumes the sunlight as energy and human consumes plant life as vitamins. The whole process of life, when it comes to the physical nature, is the cycle of copper. Humans can experience copper physically through their senses. Each one of our senses is related to an element and the nature of that sense must be fed from that element so that it can experience itself and copper is the source of this whole cycle. For example, the element Earth is experienced through the lower brain (the intestines.) It consumes copper from the foods we eat. Its nature is earth. The element air gives us the experience of copper through the nose and mouth, and through the nose and mouth we experience the smell of fragrance and the taste of flavors in foods. The element fire is experienced through upper brain and it is the mind. It operates on electrical chemical impulses.

And the last element is water. The sun is the source of light it is experienced through our (Eyes) its nature is water.

Dysfunctions of Melanocytes. There are three significant dysfunctions of melanocytes:

- ***Albinism***: is a deficit or absence of pigmentation (hypo pigmentation) caused by a mutation in the genetic encoding for tyrosinase. The body may produce little or none of this enzyme, reducing or completely blocking melanogeneisis. The dermatologic consequence is extremely light colored skin that cannot protect itself from ultraviolet light damage.

- ***Vitiligo***: is a hypo pigmentation disorder of autoimmune origin in which the melanocytes in areas of the skin die, leaving the skin without pigmentation. The cells in the skin that make up the pigment are destroyed. As a result, white patches appear on the skin in different parts of the body. Similar patches also appear on both the mucous membranes (tissues that line the inside of the mouth and nose) and the retina (inner layer of the eyeball). The hair that grows on areas affected by Vitiligo sometimes turns white. (Michael Jackson had this disease).

- ***Skin Cancer***: Melanoma is a cancer that develops in melanocytes, the pigment cells present in the *skin*. It can be more serious than the other forms of skin cancer because it may spread to other parts of the body (metastasize) and cause serious illness and death. About 50,000 new cases of melanoma are diagnosed in the United States every year. The lack of melanin can cause one to get malignant melanoma which is a serious type of skin cancer. Melanoma is often associated with death, as it causes the most numbers of casualties due to its severity. Melanoma can also be cured, but the treatment is often more extensive and evasive. Several things can contribute to the development of skin cancer. The first is prolonged exposure to the harmful UV rays of the sun. The sheer intensity of UV rays on unprotected skin raises the chances of developing skin cancer. Even with sunscreen on and clothes covering you, you can still suffer the effects from UV rays. Also, being exposed in any way to carcinogenic chemicals can increase your chances of developing skin cancer. Nicotine, arsenic, and tar are all chemicals that are harmful and can potentially lead to skin cancer. Having white skin just makes you more prone to skin cancer overall, simply because your skin will have far less melanin (the sun-protecting pigment in the skin) to combat UV rays. Those with dark skin have more melanin, which is a more natural protection and are less prone to developing skin cancer.

The skin effect and copper: The skin effect and copper is when you have a tendency of a high frequency alternating current, flowing through only the outer layer of skin. The electric current flows mainly in the skin between the outer surface and the skin depth. The skin effect causes the effective resistance of copper to increase at frequencies where the skin depth is smaller. (See Diagram A below)

The ultraviolet B light is in the color spectrum of violet. It is the fastest moving ray that the sun produces. The slowest ray is in the red spectrum, it may be classified as ultraviolet A. The ultraviolet B light is the light that does the most damage to the skin in humans, due to the fact that it moves at such a high frequency. People who have a high level of copper or melanin in their skin, do not get skin cancer from long exposer from the sun, like people who do not have as much melanin in their skin. One of the reasons why people with melanin do not get skin cancer from the sun, is because the high frequency that causes the damage rides on the outer surface of the skin (look at diagram A below), so the high frequency wave cannot penetrate the inner skin where it could cause damage to the lymphoid cells, due to the skin effect.

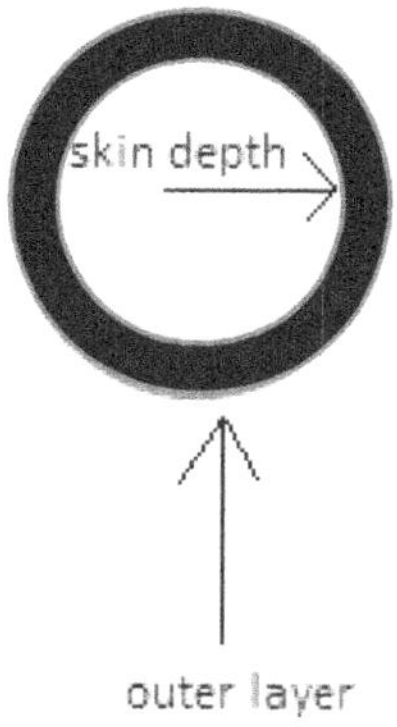

(Diagram A)

This is a perfect example in diagram A of how the pigment melanin in the skin of human life responds to the ultraviolet-light from the sun, since the source of melanin is copper.

(Diagram B)

The cross-sectional area of copper wire conducting at a high frequency, the skin depth as seen in the diagram B above. The black area means the current frequency located in the skin has penetrated deep in the surface area of the skin, (this would be the effect that the ultraviolet B- light would have on people with a low amount of melanin in their skin).

Chapter 16

Copper and the Calabash

The calabash is a system that explains the function of the sun and water. The sun is the placenta of the universe. Its principal function in the solar system is a vital organ in the universe. Without it life could not be sustained. The sun provides the fire of life in the form of light rays, which is the breath of life. The breath is moisture, which is Water and copper, they are known as the first twin births that were created according to our ancient ancestors.

We can see copper when we look at a rainbow. We can also see the variation of color in drops of water when light is reflected from it, water is a carrier of the pigment of copper it is considered copper in a wet and cool liquid form.

The sun's ultra-violet rays is a dry hot form of liquid. This dry liquid is absorbed by all life through the process of photosynthesis. So copper can be absorbed either from water or through the process of photosynthesis (which is copper being absorbed in the form of dry heat). The only difference is one is cool and wet and the other one is dry and hot.

Copper is also a fleece of vegetation, which is the essence of water and copper. Anything that can be absorb is considered a liquid.

The Calabash. The calabash is also known as the Golden Fleece. The Golden Fleece is a symbol of the ram with the sun in between his horns. The horns represents the male testicles and the sun represents the female egg, which is the fire of life which gives birth in the form of light rays. This symbol represents the perfect pair of twins in heaven (water and copper). The Golden Fleece is a reflection of light. Before the rainy season, he may be seen moving in the vault of heaven.

The ram is also a symbol of the world system, which is copper, the source material of the universe, and the creation of all life. The sun is the female which sits in between his horns, the ram body is the earth, his forehead is the moon, (a reflection of the sun), and his eyes are the stars (suns from other galaxies) in the sky. The ram's nose and mouth represent the breathing of all living creatures in the world. His hind feet are those of the large animals, his forefeet the small ones, and his tail represents the reptiles.

The Dogon Tribe says the ram is also known as the unicorn. He puts the sun (woman) on his head to hold her between his horns, (which are testicles) to have intercourse with her by means of the organ on his forehead (the unicorn). With his frontal member he fertilizes the feminine element of the sun. Water and copper, having taken the form of the ram, urinates from his lower member and showers the earth with the mists of water.

The fleece is really copper and copper is the cluster of rays coming from the rain-clouds when the sun is hidden. Copper is excreted by the water spirits in the damp atmosphere of heaven.

The copper fleece is also a fleece of light, which consists of rays. The hot dry rays are the ultra-violet light which is absorbed by all life from the sun. The cool, wet form is water. It is returned in the form of rain, both coming from the sky.

The rays of light and heat draw the water up (condensation) and also causes it to descend again in the form of rain. The movement created by this coming and going is a good thing. By means of the rays, the sun draws out water, and gives back the life-force. This movement indeed makes life.

Because fire is dry and hot, it is the cause of thirst that must be quenched through the process of condensation. We can see that the sun is giving off copper in the dry liquid form, and in return through the process of condensation quenches itself with cool wet rain water, like all thirst that needs to be fed, water is needed to sustain life. So this push pull process is a give and take process of what is being given from above is being returned from below. That which is being given is copper.

The sun is the cause of condensation causing the ram to drink the water that he is releasing. Water and copper are of the same essence and we can see their union in a rainbow. The rainbow is the symbol of the calabash in its union of dry, hot fire from the sun and cool, wet water from the earth working together as universal twins. The golden light represents the spirit of life which is the force behind all physical life, and copper is the source, although the reflection is gold. This reflection of gold light is a result of the seven different colors of light coming together. Gold is considered to be the little brother to copper because copper is the source of the reflection of the Golden Fleece.

When the ram rises to the heavens, he moves about. He goes about his business, he doesn't stay still in one place and urinate. He moves among the high clouds leaving a track of four colors from the earth as he shakes off his hooves. His left forefoot makes a black track, his right a red track, the two others one green and one yellow. The fourfold track is called the rainbow.

Chapter 17

Copper is the DNA of all Life

DNA, known as deoxyribonucleic acid, is a chemical found in the nucleus of cells and carriers the "*instruction*" for the development and functioning of living organisms. It is often compared to a set of blueprints since it contains the instructions needed to build cells. These instructions are divided into segments along a strand of DNA and are called genes. Genes are a DNA sequence that is coded for the production of a protein and control hereditary characteristics, such as eye color or personality behaviors. Proteins determine the type and function of a cell, so a cell knows whether it is a skin cell, a blood cell, a bone cell, etc... and how to perform an appropriate tasks. The DNA sequences are responsible for structural purposes or are involved in the regulation and use of genetics in animal or fungus. This is called genetic code. For example one side of DNA could have the genetic code of (Aaaa) and its complementary side would then be (Bbbb.)

The Structure: Even though the shape of DNA is often described as a ladder, it is not a straight ladder. It is twisted to the right making the shape of the DNA molecule a right handed double helix.

The DNA is a polymer (a compound, usually of high molecular weight, formed by the combination of simpler molecules (*monomers*) with millions of repeated link units. It may be formed without formation of any other product or with simultaneous elimination of water or other simple compound *(condensation p.)* The DNA has an alternating phosphate and deoxyribose units that are twisted into a double helix and are joined together by hydrogen.

DNA is Light: According to our ancient ancestors, DNA is light photons. Photons move through space as a quantum wave particle function. Atoms create their own light sphere in space-time. Light will expand out from its center in all directions, as a light sphere. This sphere moves in an ellipse spinning in a helix. (Ellipse is a closed, symmetric curve shaped like an oval, which can be formed by intersecting a cone with a plane that is not parallel or perpendicular to the cone's base), The sum of the distances of any point on an ellipse from two fixed points (called the foci) remains constant, no matter where the point is on the curve. An ellipse is a kind of stretched circle. You can conceive of grabbing the Centre of a circle and ripping the centre into two pieces, and then pulling those two portions away from each other. This would naturally, make the circle flatten at the top and the bottom and bulge at the two sides, and the two pieces of the Centre would fall along a straight line joining the two most distinct points. These two fragments of a Centre each have a name called the focus, (the focus is also known as the crossing) and the two together are the foci of the ellipse.

Everything moves in this order, even the planets. This movement takes place in one direction in a given time, and then reverses in the other direction, in another given time. This ellipse movement alternates and at the point of the crossing the polarity's of the frequencies changes.

The Fiber Skirt and magnetic field. The fiber skirt is the light photons moving in a double ellipse, weaving into a strand; these strands look like braids, this pattern moves in a helix motion. The weaving process is caused by the magnetic field. It is matter alternating into patterns that are controlled by the vibrating frequencies of the sun, planets, and stars from other galaxies etc... The double helix is known as the serpent and the crossing in the ellipse is known as the cross or crux, and the umbilical cord is the pattern of the DNA strand. Hydrogen is the element that holds the two strands together. DNA strands are copper an alkaline base, and hydrogen an acid base. Since the two are of different P.H. levels the hydrogen bond is easily broken by the magnetic field. The breaking of the bonds takes place in the middle of the strands at the point of the crux at each alternating cycle. It is at this point that the DNA is able to duplicate itself.

Cell: The DNA strand lies in the nucleus of a cell. Our very own galaxy is a cell and the center of it is where the star Sirius A is located and the nucleus is where the star Sirius B is located. It is said by our ancient ancestors that the beginning of our galaxy was formed at Sirius B. The Dogon tribe of Mali, West Africa call the Star Sirius B (Digitaria) it is the starting point of creation. It is the star that revolves around Sirius A. Digitaria is the smallest and heaviest of all the stars, it contains the germs of all things. It moves on its own axis and around Sirius A, it upholds all creation in space in this galaxy.

The orbit of Digitaria is egg-shaped or elliptical and has one of its main stars Sirius A, as one of the foci (crossing points) of its orbit. This movement is the same movement of our DNA.

Neuron: A neuron is a cell that processes information in the brain. It passes electrical chemical messages among themselves and they are connected in large complex networks. Link to our senses, they register what is happening in the world around us, and linked to our muscles they allow movement and action. The neuron has certain core components. It has a protein skeleton (*are color strands of copper*) that holds its shape. The outer "skin" of a neuron is called a cell membrane, it is made of fluid and fatty material. Some proteins are stuck into this membrane, which means they have access to the outside of the neuron. Since the membrane is fatty, chemicals that dissolve well in fat, such as the hormones testosterone and estrogen can pass right through it. Neurons also have many organelles, and analogous to miniature internal organs. The nucleus is an organelle that contains all of our DNA, or a complete genetic blueprint.

From this blueprint, the neuron can make anything that it needs. Mitochondria are the other organelles that are needed to make energy. They take oxygen and sugar glucose, (delivered by the blood supply to the brain) and convert them into ATP *(adenosine triphosphate).*

Parts of the Neuron: Dendrites (meaning branches of the tree) extend from the body of a neuron. These branches are typically the receiving end, taking in messages from the other neurons. In fact, whenever the brain learns new information, the shapes of the dendrites change, rewriting, in a sense to code the new information.

Cell body: The cell body, or soma, integrates all of the input signals from other neurons. Neurons either excite or inhibit each other, so the soma must determine which influences is greater. The soma also contains the nucleus of the neuron, where genetic instructions are stored and copied to make proteins.

Axons: The axon is a long arm extending from the soma. When excitation from inputs outweighs inhibition, a message is sent down the axon (called an action potential), that in turn, reaches other neurons. The axon hillock is the transition of the soma to the axon, and it is this part that is critical to deciding whether or not a message is sent. When the action potential reaches the end of the axon, chemical messages call *neurotransmitters* are released.

Electrical Transmission: Neurons have a small voltage, which allows them to send electrical impulses. The neuron's voltage is created by an imbalance of charge particles called ions. When at rest, the neuron's charge is relatively more negative, this is called its resting potential.

However, when excited by other neurons, certain ions enter the neuron and make it more positive. If the charge is sufficient, the neuron sends an electrical signal, the action potential. The action potential spreads along the axon, reaching the end, where it connects with other neurons. Afterward, the ions return to their original positions and the neuron returns to a negative resting state.

DNA Chain: Cells are light partials, so is the DNA chain. It is a strand of color patterns in a sequence of numbers. And based on the sequence of the number, you can determine what the creation is going to be.

The Italian mathematician "Leonardo Fibonacci" (can. 1175-1250) discovered a spiral related sequence of numbers, the Fibonacci numbers while studying the birth pattern of rabbits. Through later discoveries, it was found to mathematically explain many of the patterns underlying structures in nature. The Fibonacci series or numbers are 0; 1; 1; 2; 3; 5; 8; 13; 21; etc... Studies have shown that the center of a daisy is intersected by a set of spirals turning in a clockwise direction and another anti clockwise.

The number of spirals in each set is a Fibonacci number, and so is the number of petals. The number of leaves of plants, when counted following spiraling motion, as well as the seeds of sunflowers and daisies, the scales of pineapples and pine cones all add up to a Fibonacci number. Even the distance between leaves, which is usually a fraction of a full turn of a spiral away, yield meaningful fractional numbers (*phyllotaxis.*) There are specific to each species.

If you take the number patterns of Fibonacci and add colors to the numbers, you will come up with the DNA strand. Every living creature has its own DNA pattern. This pattern is made of light, and the sun is the source of light, so the sun is the source of DNA patterns. The light photons that is coming from the sun changes as the sun moves in the universal sky. These light waves are measured in angles every nanosecond and at each nanosecond you have one of the color patterns in the spectrum dominating. The dominating colors resonate with all the same colors throughout the universe causing that color to move at that given time. This unforeseen dance creates patterns of light that vibrate at a given frequency. These frequencies changes the patterns.

The DNA of the Sun. In humans a chromosome is an organized structure of DNA. It is similar to the sun's DNA which lies in the *Chromosphere* (literally, "*sphere of color*"). It is the second of the three main layers in the sun's sphere. It lies in the middle, and just like in the human DNA, hydrogen bonds hold the light particles together in the chromosphere.

The creator is a cosmic mind that holds all things together and the source it chose to use as its building material is copper. The Dogon tribe of Mali, West Africa consider copper to be impure fire, i.e., fire which has undergone some mutation. In its second stage of fire it is "solid" fire as an element, it is the source of creation. Like the human placenta, which will always remain "alive", even after coming out of the womb, copper also called "copper of past times", is the symbol of permanence because "it does not die".

This means copper is the DNA of all life and it can mutate and reproduce itself.

Chapter 18

Copper is the Life Source for the Midbrain and Pineal Gland

The Midbrain. The midbrain, also called mesencephalon, is the region of the developing vertebrate brain containing cranial nerves that stimulate the muscles controlling eye movement, lens shape, and pupil diameter. These masses of nerves form the nuclear complex of the oculomotor nerve and the trochlear nucleus. The midbrain is located between the two developmental regions of the brain known as the forebrain and hindbrain. Within the midbrain is the reticular formation, which is part of the tegmentum, a region of the brainstem that influences motor functions. The midbrain also contains the crus cerebri, which is made up of nerve fibers connecting the cerebral hemispheres to the cerebellum, and a large pigmented nucleus called the *substantia nigra*.

The substantia nigra consists of two parts, the pars reticulata and the pars compacta. Cells of the pars compacta contain the dark pigment *melanin*; these cells synthesize *dopamine* and project to either the *caudate nucleus* or the *putamen*, both of which are structures of the basal ganglia and are involved in mandating movement and coordination. These two structures, in addition to the globus pallidus, form the striatum. By inhibiting the action of *neurons* in the caudate nucleus and the putamen, the dopaminergic cells of the pars compacta influence the neuronal output of the *neurotransmitter GABA (gamma-aminobutyric acid).* The neurons, in turn, project to the cells of the pars reticulata, by projecting fibers to the thalamus, which are a part of the output system of the corpus striatum.

At the caudal (rear) midbrain, crossed fibers of the superior cerebellar peduncle (the major output system of the cerebellum, also known as the hindbrain) surround and partially terminate in a large centrally located structure known as the *red nucleus.* Most crossed ascending fibers of this bundle project to thalamic nuclei, which have access to the primary motor cortex. A smaller number of fiber synapsis on large cells in the caudal regions of the red nucleus; these synapsis give rise to the crossed fibers of the rubrospinal tract, which runs to the *spinal cord* and is influenced by the motor cortex.

The roof plate of the midbrain is formed by two paired rounded swellings, the superior and inferior colliculi. The *superior colliculus* receives input from the *retina* and the visual cortex and participates in a variety of visual reflexes, particularly the tracking of objects in the visual field. The *inferior colliculus* receives both crossed and uncrossed auditory fibers and projects upon the medial geniculate body, the auditory relay nucleus of the thalamus.

Pigments in the brain. *Dopamine;* is found in the frontal brain, *Substantia Nigra;* is found deep in the mid brain, *Gamma- Aminobutyric Acid* also known as the *Red Nucleus;* is found in the mid brain as well, *Norepinephrine;* is found in the locus coeruleus also known as the Blue Place, *Melanopsin*; In humans, melanopsin is found in intrinsically photosensitive retinal ganglion cells, *Gray Matter*; found in the spinal fluid and *Dipropyltryptamine; Also known as light found in the pineal gland.*

The Pineal Gland: The pineal gland is an endocrine gland in the vertebrate brain. The shape resembles a tiny pine cone and it is located near the center of the brain, between the two hemispheres known as the valley, tucked in a grove where the two round thalamic bodies join.

The pineal gland is an electrical source that sends and receives its signals from the vibrating rhythms of the universe. The pineal gland synthesizes and secretes copper and water, which is the source of light that communicates information to various parts of the body.

Ultimately, copper is the source of melatonin, which has the ability to entrain biological rhythms and has the important effects on the reproductive function of many animals, as well as humans.

The light transducing ability of the pineal gland has led some to call the pineal gland the third eye, which is intuitive by nature, but the third eye is not the pineal gland alone, it is the whole region of the midbrain. This region of the midbrain must be flooded with water or spinal fluid in order for the third eye to see. The brain is the upper hemisphere of the body, it is electrical. In order for the brain to be electrical, it has to filter out all matter. It does this through a process called *the blood-brain barrier*. The barrier consists of extra small pores in the blood vessels that supply the brain. In most of the body, the pores are large and most things can pass through. But in the brain, the pores are extremely small, so that only very small molecules (such as water) can pass through. Remember water and copper are one and the same.

The pineal gland is the source of melatonin a hormone derived from tryptophan that regulates circadian rhythm (sleep cycle.) The pineal gland develops from the roof of the diencephalon, a section of the brain. And in some lower vertebrates the gland has a well-developed eyelike structure; and others, though not organized as an eye, it functions as a light receptor.

Serotonin is a neurotransmitter in the nervous system and melatonin varies in a daily cycle, thereby allowing the entrainment of the circadian rhythms of several biological functions, they are both being modulated. Modulation deals with aspects frequencies, waveform, carrier signals, and amplitude phase.

Serotonin is a process of photosynthesis which takes place during the daytime. It capture the sun light and stores it in the pigment cells.

Melatonin is the process that takes place during the night time hours. It is when the captured sun light pigment cells become activated at night, around 9p.m. to 4a.m. The peak hour is about 12a.m. These light cells start repairing the body, killing off the old cells and replacing them with the new light cells, so that growth can take place.

If the body does not get the proper rest, it will hinder the melatonin process and the old cells will not be replaced by the newer ones and the older ones will become cancerous.

Biological Rhythms. Light is the source of the biological rhythms that are in synchrony with the circadian rhythms of nature. Circadian rhythms are closely linked with the movements of our planet around the Sun. They are also influenced by the motions of the Moon and the other planets in relation to the position of the earth. Our body follows more than a thousand 24-hour, hour rhythms. Each individual rhythm controls the timing of an aspect of our body functions, including heart rate, blood pressure, body temperature, hormone levels, secretion of digestive juices, and even pain threshold. All these rhythms are well coordinated with one another and controlled by the brain's pacemaker device, known as suprachiasmatic nuclei.

This area of the brain regulates the firing of nerve cells that seems to set the clock of our biological rhythms. If one rhythm is somewhat disrupted, other rhythms are thrown off balance too. In fact numerous disorders can arise from interference with one or more of our biological rhythms because of an unbalanced irregular lifestyle.

Suprachiasmatic Nuclei. (SCN) the suprachiasmatic nuclei is a small, paired, winged shaped structure (the pair of wings in the kundalini symbol) in the hypothalamus located at the base of the brain. Within each side of the SCN is a network up to several thousand neurons. Individual isolated SCN neurons suggest that each SCN cell is a functional clock, normally synchronized with the activity of its neighbors.

Light enters the eye and activates neurons in the retina that convert photons (light particles) to electrical signals. The retinal neurons transmit the electrical signals from the retina via long axons in the optic nerve, along the way is the optic chiasm. Visual information continues toward the back of the brain, where it is processed into images that we can consciously perceive.

The neurons carrying information to the SCN however, take a different path. They exit the optic chiasm and turn upward, toward the SCN. (Suprachiasmatic means above the chiasm.) Circadian rhythms can be entrained (the synchronization and control of cardiac rhythm by an external stimulus) by the light and dark cycle of the environment. Once light coming from the environment reaches the eyes, specialized light sensitive to retinal cells, known as photoreceptors convert the light into electrical signals then transmit it to the brain. These signals that travel along the retinohypothalamic tract produce entrainment. Entrainment is shifting the phase of the suprachiasmatic nucleus activity, so that it synchronizes with the light and dark cycle of the environment of the same frequency. During subjective daytime, entrainment is mandated by cyclic adenosine monophosphate and pituitary adenyl cyclase-activating peptide, whereas during subjective nighttime, it is mandated by cyclic guanosine monophosphate and acetylcholine. As mentioned, the SCN receives its inputs from the human eyes, in particular from the photosensitive ganglion cells in the retina. These ganglion cells contain a chemical call melanopsin which is sensitive to the blue region, a visual light frequency.

SCN has two parts, dorsal and ventral. The dorsal has a 24 hour period regardless of light input to the eyes, thus it will continue to work even in constant darkness. The ventral gets the input from the eyes and thus distinguishes night from day. Thus by giving eyes light during the night regularly and depriving light during the day. The ventricle can be trained to reverse its cycle. In chronobiology this period matching training exercise of the ventral part of the SCN is called entrainment. After entrainment, the cycle persist and to change it again the inputs need to be reversed for sufficient time. Improper entrainment can cause health problems.

Human SCN Anatomy Background. Aside from the environmental light signals, the suprachiasmatic nuclei is influence by melatonin. Melatonin is a hormone synthesized from the neurotransmitter, serotonin is synthesized by the pineal gland. It is both an agent and regulator of the suprachiasmatic nucleus. Its synthesis is circadian, wherein it is inhibited by light and stimulated by darkness. As its level increases during environmental darkness, melatonin sends information to the brain that night time has arrived, and the body's functions entrainment is evoked accordingly.

Living organisms have evolved internal timekeeping mechanisms to synchronize behavior and physiology with the cycles of day and night.

These biological clocks have been found in organisms as diverse as fungi, fruit flies, hamsters, and humans. The biological clock of humans is found deep within the brain. Several genes has been identified that regulate the circadian rhythm. These are the timeless, period, clock, and cycle genes. They control the protein expression leading to a complex intercellular negative feedback system.

The negative feedback system accounts for the periodicity of the suprachiasmatic nucleus. Variation in the rhythm of neuronal and hormonal signaling occurs seasonally. It may reflect the influence of variation in the length, tides, temperature, and sunshine of daylight. In jet lag, the biological clock is disrupted due to a sudden artificial change in clock time from traveling to a different time zone, resulting in an abrupt shift of the neuronal and hormonal signals. Oftentimes, it requires several days to reset back the body functioning within a normal 24-hour cycle.

The Awakening of the Third Eye through Meditation. There are several things that must take place before the third eye can be activated; that is to shut down the *Parasympathetic nervous system.* Sometimes called the rest and digest system, the parasympathetic nervous system conserves energy as it slows the heart rate, increases intestinal and gland activity, and relaxes sphincter muscles in the gastrointestinal tract. The parasympathetic nervous system runs along the spinal cord. The spinal energy moves outward and downward like branches on a tree and at the end of the branches are the different organs that the nerves are connected to. These organs are consuming the energy the spinal cord is providing, and it is this energy that must be rerouted inward and upward into the brain.

Meditation. When meditating the posture is important. The spine should be straighten, and the neck should also be held straight. The shoulders should be pushed back and the chest forward with the abdomen in. A bent spine throws the vertebrae out of their proper order. This would kill the spinal energy from reaching the brain when the spinal energy reverses upwards. The biological clock has to be alerted so that the pigments can be secreted. Your attention should be placed in the middle of the forehead. This is the most concentrated pathway of *Dopamine* (Dopaminergic Pathways, sometimes called Dopaminergic Projection). The Dopaminergic travels in the nerves cells in such areas as the *substantia nigra,* which acts as a booster and mixes with all the other pigments. The substantia nigra and dopamine will flow when the body goes into a rest state. This will cause the body to become paralyzed and the mind would lose consciousness of the physical body. The body goes through this state every night when we go to sleep, although we are unaware of this process taking place. This is because we lose consciousness of self. The mind, in a sense, is being trick into believing it is time to go to sleep. This is all controlled in the midbrain area of the basal ganglia, where the thalamus, and the suprachiasmatic nuclei sits.

When the body goes into a state of paralysis, there are three spinal centers that should be concentrated on, the coccyx, dorsal, and medulla centers. These are important vantage points in which the consciousness becomes ensconced while moving upward through the centers of the brain. There is a special polarity between these centers that helps the attuned consciousness to be lifted upward.

First there is the magnetism between the negative pole of the coccyx center and the higher positive dorsal center. By deep meditation, when the consciousness is raised to the dorsal center, that center becomes the negative pole and the medulla center becomes the positive pole, pulling the consciousness upward to the center of the highest realization of the brain. This upward moving energy causes the coccyx bone to vibrate, which helps the movement of the current flow. (If you ever experienced riding down a roller coaster that funny feeling you get in your lower stomach is caused by the vibration of the coccyx bone.) This upward movement of energy is known as the kundalini. Once the kundalini energy moves upward into the pons varolii (a part of the brain stem situated above the medulla which is centered below the two hemispheres of the cerebrum), it contains the ascending sensory and descending motor tracts that connect the brain to the rest of the body. These tracts travel through a network of nerve cells, called the reticular formation, whose function is to arouse activity to the rest of the brain and to regulate the twenty- four-hour cycle of sleep and waking state. The pons varolii contains a particular structure, the *Locus Coeruleus ("The Blue Place")* a small, concentrated cluster of cells containing norepinephrine, a chemical substance that stimulates the mobilization and prepares the body for action. This structure is involved in the arousal, dreaming, sleep, and mood state.

Once the kundalini energy reaches the locus Coeruleus, the hormones from the blue place and the hormones from the substantia nigra are mixed. The black pigment act as an amplifier, when mixed with the blue pigment, it heightens the electrical current moving in the medulla pons area. The amplified current flow moves into the cerebellum where it takes root. The current then leaves the cerebellum and enters the *red nucleus* where it mixes with the substantia nigra.

The high current causes the spinal fluid *(Gray Matter)* to move upward into the brain where the fluid rushes into the thalamus, (which lies in the center of the midbrain along the valley). From there it rushes into the midbrain, known as the third eye. This process of the third eye being flooded with water must take place in order to see spiritual light. The fluid rush causes the pineal gland to vibrate and to secrete its own fluid "Dipropyltryptamine". The vibration of the pineal gland causes the roof plate to vibrate, raising the consciousness to a super consciousness level which is *(light).* This energy vibrates on the universal frequency. The vibration spreads throughout the brain freeing the spirit from the earthly body, known as the kundalini energy taking flight.

Closing Remarks

I am only scratching the surface when it comes to the role that copper plays in all things. Copper is essential to all life when it comes to their natural functions. The function of the universe is similar to the function of the body. The sun is a major organ of the universe and it produces copper in the dry, liquid state of heat rays. The two major organs in the body are the liver and the brain, they are the main sources of copper production. The liver provides copper in the form of bile, which is made from cholesterol, and the brain is 75% cholesterol, the fatty cells in the body is what make up cholesterol cells and light. The sun pigment source is given off in the form of ultraviolet light rays, the ultraviolet light rays are absorbed and stored in the fatty cells of the body. The captured light rays from the sun lies dormant in the day time, this is the process of serotonin. At night fall, around 9 o'clock p.m. to midnight, are the peak hours when the light cells go to work in replacing the old dead cells with new rejuvenating body cells to stay alive; this process is known as melatonin. The light cells move with the rhythm of the universe. This light energy is what makes up all of your steroid hormones androgen, estrogen, progesterone and testosterone. These hormones are just a combination of fatty cells and copper or light. The whole process of the fatty cells storing this light energy in the day time and releasing it at night is the process of photosynthesis. The brain is made of fatty cells and light. The DNA is made of a strand of light patterns and so is the spiritual nature. The very essence of life is light, which make us light beings and if we are light beings then we are copper beings, because copper is the source of light. Our bone joints are a very important source of copper production, they produce copper in the form of semen. The semen produces the placenta and all placentas are made of copper.

There are a few books that I would like to mention, they were essential in my research; the first book is, "*Conversations with Ogotemmeli an Introduction to Dogon Religious Ideas"* by Marcel Griaule. This was the book that opened my eyes. It gave me the understanding of what material source were used by the Almighty Creator for His creations; which is copper. When I found out that Marcel Griaule wrote another book on the Dogon Tribe "*The Pale Fox*", by M. Griaule & G. Dieterlen, I had to buy it. At that time, the book was priced at $100.00. Needless to say, I bought the book and don't regret it to this day; because this book changed my life. It took me a while to understand a great deal of the material in the book, so I took the book with me wherever I went, as if it was the Holy Bible. In fact, it is my bible. I don't let a day go by without reading it. Every time I open the book, I learn something new about the creation of life. This is the type of book you would have to read daily in order to understand it. The next book I would like to mention is "*The Amazing Liver & Gallbladder Flush", by* Andreas Moritz. This is one book that can help save your life if you read it. The information in this book shows you how to remove the gallstones from your liver and gallbladder. I have done it at least seven times now and I am very happy with the results. Some of the information in my book concerning the liver was taken from his book.

Glossary

CHAPTER 2

Acid: (ăs′ĭd) ***Chemistry*** any of a class of substances whose aqueous solutions are characterized by a sour taste, the ability to turn blue litmus red, and the ability to react with bases and certain metals to form salts; having a pH of less than 7; having a relatively high concentration of hydrogen ions.

Ammonia: (əˈməʊnɪə; -njə) (Elements & Compounds) a colorless pungent highly soluble gas mainly used in the manufacture of fertilizers, nitric acid, and other nitrogenous compounds, and as a refrigerant and solvent. (Elements & Compounds) a solution of ammonia in water, containing the compound ammonium hydroxide.

Base: A base is an alkaline

Fucoxanthin: (fyo͞o′kō-zăn′thĭn) a brown or olive- green carotenoid pigment, C40H60O6, found in brown alga.

Hydroscope: (hī′drə-skōp′).*a*n optical device used for viewing objects far below the surface of water.

Hydroxide: (Elements & Compounds) a base or alkali containing the ion OH (Elements & Compounds) any compound containing an –OH group.

Malleability: ***Ductility:*** a solid material's ability to deform under tensile stress; this is often characterized by the material's ability to be stretched into a wire., a similar property, is a material's ability to deform under compressive stress; this is often characterized by the material's ability to form a thin sheet by hammering or rolling. Both of these mechanical properties are aspects of plasticity, the extent to which a solid material can be plastically deformed without fracture.

Phaeophytin or **phaeophytin:** (abbreviated Pheo) a chemical compound that serves as the first electron carrier intermediate in the electron transfer pathway of photosystem II (PS II) in plants, and the photosynthetic reaction center (RC P870) found in purple bacteria. In both PS II and RC P870, light drives electrons from the reaction center through phaeophytin, which then passes the electrons to a Quinone (QA) in RC P870 and RC P680.

Potassium Ferrocyanide: (Elements & Compounds) a yellow soluble crystalline compound used in case- hardening steel and making dyes and pigments. Formula: K4Fe (CN) 6. Also called: potassium hexacyanoferrate (II) *or* yellow prussiate of potash.

Potassium hydroxide: an inorganic compound with the formula KOH, commonly called caustic potash. Along with sodium hydroxide (NaOH), this colorless solid is a prototypical strong base.

Sulfuric acid: a clear, colorless to brownish, dense, oily, corrosive, water-miscible liquid, H2SO4, used chiefly in the manufacture of fertilizers, chemicals, explosives, and dyestuffs and in petroleum refining. Also called oil of vitriol.

CHAPTER 3

Aerobic: "requiring air", where "air" usually means oxygen.

Anthocyanins: [also anthocyans; from Greek *(anthos)* flower + *(kyanos)* = blue] are water soluble vacuolar pigments that may appear red, purple, or blue depending on the ph. They belong to a parent class of molecules called flavonoids synthesized via the phenylpropanoid pathway; they are odorless and nearly flavorless, contributing to taste as a moderately astringent sensation. Anthocyanins occur in all tissues of higher plants, including leaves, stems, roots, flowers, and fruits. Anthoxanthins are clear, white to yellow counterparts of anthocyanin's occurring in plants. Anthocyanins are derivatives of anthocyanin's, which include pendant sugars.

Beryllium: the chemical element with the symbol (Be) and atomic number 4. Because any beryllium synthesized in stars is short-lived, it is a relatively rare element in both the universe and in the crust of the Earth. It is a divalent element which occurs naturally only in combination with other elements in minerals. Notable gemstones which contain beryllium include beryl (aquamarine, emerald) and chrysoberyl. As a free element it is a steel gray, strong, lightweight and brittle alkaline earth metal.

Bismuth: a chemical element with symbol (Bi) and atomic number 83. Bismuth, a prevalent poor metal, chemically resembles arsenic and antimony. Elemental bismuth may occur naturally, although its sulfide and oxide form important commercial ores. The free element is 86% as dense as lead. It is a brittle metal with a silvery white color when freshly produced, but is often seen in air with a pink tinge owing to surface oxidation. Bismuth is the most naturally diamagnetic and has one of the lowest values of thermal conductivity among metals.

Calcium: is the chemical element with symbol (Ca) and atomic number 20. Calcium is a soft gray alkaline earth metal, and is the fifth most abundant element by mass in the Earth's crust. Calcium is also the fifth most abundant dissolved ion in seawater by both molarity and mass, after sodium, chloride, magnesium, and sulfate.

Carotenoids: organic pigments that are found in the chloroplasts and chloroplasts of plants and some other photosynthetic organisms like algae, some bacteria, and some fungi. Carotenoids can be produced from fats and other basic organic metabolic building blocks by all these organisms. Carotenoids generally cannot be manufactured by species in the animal kingdom so animals obtain carotenoids in their diets, and may employ them in various ways in metabolism.

Chlorophyll: (also chlorophyl) a green pigment found in cyanobacteria and the chloroplasts of algae and plants. Its name is derived from the Greek words, *chloros* ("green") and, *phyllon* ("leaf"). Chlorophyll is an extremely important biomolecule, critical in photosynthesis, which allows plants to absorb energy from light. Chlorophyll absorbs light most strongly in the blue portion of the electromagnetic spectrum, followed by the red portion. However, it is a poor absorber of green and near-green portions of the spectrum, hence the green color of chlorophyll-containing tissues.

Corundum: a crystalline form of aluminum oxide (Al2O) with traces of iron, titanium and chromium. It is a rock forming mineral. It is one of the naturally clear transparent materials, but can have different colors when impurities are present. Transparent specimens are used as gems, called ruby if red and padparadscha if pink-orange. All other colors are called sapphire, "green sapphire" for a green specimen.

Cyanobacteria: also known as blue-green bacteria, blue-green algae, and Cyanophyta; is a phylum of bacteria that obtain their energy through photosynthesis. The name "cyanobacteria" comes from the color of the bacteria (Greek: (kyanos) = blue.

Cysteine: (abbreviated as Cys or C) α-amino acid with the chemical formula HO2CCH (NH2) CH2SH. It is a semi-essential amino acid, which means that it can be biosynthesized in humans.

Ferric: refers to iron containing materials or compounds. In chemistry the term is reserved for iron with an oxidation number of

+3, also denoted iron (III) or Fe^{3+}. On the other hand, ferrous refers to iron with oxidation number of +2, denoted iron

(II) or Fe^{2+}. Iron (III) is usually the most stable form of iron in air, as illustrated by the pervasiveness of rust, an insoluble iron (III) containing material.

Ferromagnetism: the basic mechanism by which certain materials (such as iron) form permanent magnets, or are attracted to magnets. In physics, several different types of magnetism are distinguished. Ferromagnetism (including ferrimagnetism) is the strongest type; it is the only type that creates forces strong enough to be felt, and is responsible for the common phenomena of magnetism encountered in everyday life. Other substances respond weakly to magnetic fields with two other types of magnetism, paramagnetism and diamagnetism, but the forces are so weak that they can only be detected by sensitive instruments in a laboratory. An everyday example of ferromagnetism is a refrigerator magnet used to hold notes on a refrigerator door. The attraction between a magnet and ferromagnetic material is "the quality of magnetism first apparent to the ancient world, and to us today".

Glutathione: (**GSH**) a tripeptide with a gamma peptide linkage between the amine group of cysteine (which is attached by normal peptide linkage to a glycine) and the carboxyl group of the glutamate side-chain. It is an antioxidant, preventing damage to important cellular components caused by reactive oxygen species such as free radicals and peroxides.

Halite: (hælaɪt) commonly known as rock salt, the mineral form of sodium chloride (NaCl). Halite forms isometric crystals. The mineral is typically colorless or white, but may also be light blue, dark blue, purple, pink, red, orange, yellow or gray depending on the amount and type of impurities. It commonly occurs with other evaporated deposit minerals such as several of the sulfates, halides, and borates.

Hemoglobin: abbreviated **(Hb)** or **(Hgb)**, is the iron containing oxygen transport metalloprotein in the red blood cells of all vertebrates (with the exception of the fish family Channichthyidae as well as the tissues of some invertebrates.
Hemoglobin in the blood carries oxygen from the respiratory organs (lungs or gills) to the rest of the body (i.e. the tissues) where it releases the oxygen to burn nutrients to provide energy to power the functions of the organism, and collects the resultant carbon dioxide to bring it back to the respiratory organs to be dispensed from the organism.

Lime: white, lumpy, caustic powder made of calcium oxide sometimes mixed with other chemicals. It is made industrially by heating limestone, bones, or shells. Lime is used as an industrial alkali, in waste treatment, and in making glass, paper, steel, insecticides, and building plaster. It is also added to soil to lower its acidity.

Magnesium: a chemical element with the symbol (Mg) and atomic number 12 it's common Oxidation number is +2. It is an alkaline earth metal and the eighth most abundant element in the Earth's crust and ninth in the known universe as a whole. Magnesium is the fourth most common element in the Earth as a whole (behind iron, oxygen and silicon), making up 13% of the planet's mass and a large fraction of the planet's mantle.
The relative abundance of magnesium is related to the fact that it easily builds up in supernova stars from a sequential addition of three helium nuclei to carbon (which in turn is made from three helium nuclei). Due to magnesium ion's high solubility in water, it is the third most abundant element dissolved in seawater.

Magnetite: a mineral, one of the two common naturally occurring iron oxides (chemical formula Fe3O4) and a member of the spinel group. Magnetite is the most magnetic of all the naturally occurring minerals on Earth. Naturally magnetized pieces of magnetite, called lodestone, will attract small pieces of iron, and this was how ancient people first noticed the property of magnetism.

Melanin: (mɛlənɪn) (Greek: *melas*, "black") a ubiquitous natural pigment found in most organisms (spiders are one of the few groups in which it has not been detected). In animals, melanin pigments are derivatives of the amino acid tyrosine. There are three basic types, namely eumelanin, pheomelanin, and neuromelanin. The most common biological type is eumelanin, which is a brown-black polymer of dihydroxyindole carboxylic acids and their reduced forms. Pheomelanin is a cysteine-containing red-brown polymer of benzothiazine units largely responsible for red hair and freckles. Neuromelanin is present in the brain, but its function remains obscure.

Niobium: formerly columbium; a chemical element with the symbol (Nb) (formerly Cb) and atomic number 41. It is a soft, grey, ductile transition metal, which is often found in the pyrochlore mineral, the main commercial source for niobium, and columbite. The name comes from Greek mythology. Niobe, daughter of Tantalus.

Oxide: a chemical compound that contains at least one oxygen atom and one other element in its chemical formula. Metal oxides typically contain an anion of oxygen in the oxidation state. Most of the Earth's crust consists of solid oxides, the result of elements being oxidized by the oxygen in air or in water.

Phycobilin: (from Greek: *(phykos)* meaning "alga", and from Latin: *bills* meaning "bile") chromophores (light-capturing molecules) found in cyanobacteria and in the chloroplasts of red algae, glaucophytes and some cryptomonads (though not in green algae and higher plants). They are unique among the photosynthetic pigments in that they are bonded to certain water-soluble proteins, known as Phycobiliproteins. Phycobiliproteins then pass the light energy to chlorophylls for photosynthesis.

Talc: (derived from Persian: *tālk*; Arabic: *talk*) a mineral composed of hydrated magnesium silicate with the chemical formula $H_2Mg_3(SiO_3)_4$ or $Mg_3Si_4O_{10}(OH)_2$. In loose form, it is the widely used substance known as talcum powder. It occurs as foliated to fibrous masses, and in an exceptionally rare crystal form. Talc is not soluble in water, but it is slightly soluble in dilute mineral acids. Its color ranges from white to grey or green and it has a distinctly greasy feel. Its streak is white.

Titanium: a chemical element with the symbol (Ti) and atomic number 22; a lustrous transition metal with a silver color, low density and high strength. It is highly resistant to corrosion in sea water, aqua regia and chlorine.

Tyrosine: abbreviated as (**Tyr** or **Y**) or 4- hydroxyphenylalanine, one of the 22 amino acids that are used by cells to synthesize proteins. Its codons are UAC and UAU. It is a non-essential amino acid with a polar side group. The word "tyrosine" is from the Greek *tyri*, meaning cheese, as it was first discovered in 1846 by German chemist Justus von Liebig in the protein casein from cheese. It is called tyrosyl when referred to as a functional group or side chain.

Tungstate: a compound that contains an oxyanion of tungsten or is a mixed oxide containing tungsten. The simplest tungstate is salt of any tungstic acid ion.

Yttrium: a chemical element with symbol (Y) and atomic number 39. It is a silvery-metallic transition metal chemically similar to the lanthanides and it has often been classified as a "rare earth element". Yttrium is almost always found combined with the lanthanides in rare earth minerals and is never found in nature as a free element.

CHAPTER 4

Antiferromagnetic: materials occur commonly among transition metal compounds, especially oxides. Examples include hematite, metals such as chromium, alloys such as iron manganese (FeMn), and oxides such as nickel oxide (NiO). There are also numerous examples among high nuclearity metal clusters. Organic molecules can also exhibit antiferromagnetic coupling under rare circumstances, as seen in radicals such as 5- dehydro-m-xylylene.

Aura: an elemental spirit in the *Seiken Densetsu* Series.

Diamagnetism: the property of an object or material that causes it to create a magnetic field in opposition to an externally applied magnetic field. It is a quantum mechanical effect that occurs in all materials; where it is the only contribution to the magnetism the material is called a *diamagnet*.

Unlike a ferromagnet, a diamagnet is not a permanent magnet. Its magnetic permeability is less than $\mu 0$ (the permeability of free space). In most materials diamagnetism is a weak effect, but a superconductor repels the magnetic field entirely, apart from a thin layer at the surface.

Electromagnetic: pertaining to or exhibiting magnetism produced by electric charge in motion; "electromagnetic energy".

Paramagnetism: a form of magnetism whereby certain materials are attracted by an externally applied magnetic field. In contrast with this behavior, diamagnetic materials are repelled by magnetic fields. Paramagnetic materials include most chemical elements and some compounds they have a relative magnetic permeability greater or equal to 1 (i.e., a positive magnetic susceptibility) and hence are attracted to magnetic fields. The magnetic moment induced by the applied field is linear in the field strength and rather weak. It typically requires a sensitive analytical balance to detect the effect and modern measurements on paramagnetic materials are often conducted with a SQUID magnetometer.

CHAPTER 5

Consciousness: the state or condition of being conscious. A sense of one's personal or collective identity, including the attitudes, beliefs, and sensitivities held by or considered characteristic of an individual or group.

Fetuses: the unborn young of a viviparous vertebrate having a basic structural resemblance to the adult animal. In humans, the unborn young from the end of the eighth week after conception to the moment of birth, as distinguished from the earlier embryo.

Oviduct: a tube through which the ova pass from the ovary to the uterus or to the outside.

Spectrum: Physics: the distribution of a characteristic of a physical system or phenomenon, especially the distribution of energy emitted by a radiant source, as by an incandescent body, arranged in order of wavelengths. The distribution of atomic or subatomic particles in a system, as in a magnetically resolved molecular beam, arranged in order of masses.

CHAPTER 6

Conceptus: derivatives of zygote denotes the embryo and its adnexa (appendages or adjunct parts) or associated membranes (i.e. the products of conception). The conceptus includes all structures that develop from the zygote, both embryonic and extraembryonic. It includes the embryo as well as the embryonic part of the placenta and its associated membranes amnion, chorion (gestational sac), and yolk sac.

Cytokines: (Greek *cyto-*, cell; and *-kinos*, movement) small signaling molecules used for cell signaling. Cytokines can be classified as proteins, peptides, or glycoproteins; the term "cytokine" encompasses a large and diverse family of regulators produced throughout the body by cells of diverse embryological origin.

Endometrium: the innermost glandular layer and functions as a lining for the uterus, preventing adhesions between the opposed walls of the myometrium, thereby maintaining the patency of the uterine cavity. During the menstrual cycle or estrous cycle, the endometrium grows to a thick, blood vessel rich, glandular tissue layer.

Estriol: estrogenic hormone, $C_{18}H_{24}O_3$, found in the urine during pregnancy. Also called *theelol* oxytocin. A short polypeptide hormone, $C_{43}H_{66}N_{12}O_{12}S_2$, released from the posterior lobe of the pituitary gland that stimulates the contraction of smooth muscle of the uterus during labor and facilitates ejection of milk from the breast during nursing.

Follicle stimulating hormone: a gonadotropic hormone of the anterior pituitary gland that stimulates the growth of follicles in the ovary and induces the formation of sperm in the testis (FSH).

Gonadotropin: a hormone that stimulates the growth and activity of the gonads, especially any of several pituitary hormones that stimulate the function of the ovaries and testes.

Myometrium: the middle layer of the uterine wall, consisting mainly of uterine smooth muscle cells (also called uterine myocytes, but also of supporting stromal and vascular tissue. Its main function is to induce uterine contractions.

Nebula: (from Latin: cloud. nebulae or *nebulæ*, with ligature or *nebulas*) an interstellar color clouds of dust, hydrogen, helium and other ionized gases. Originally, *nebula* was a name for any diffuse astronomical object, including galaxies beyond the Milky Way.

Placenta: an organ that connects the developing fetus to the uterine wall to allow nutrient uptake, waste elimination, and gas exchange via the mother's blood supply.

Progestin: a synthetic progestogen that has progestational effects similar to progesterone. The two most common uses of progestins are for hormonal contraception (either alone or with an estrogen), and to prevent endometrial hyperplasia from unopposed estrogen in hormone replacement therapy. Progestins are also used to treat secondary amenorrhea, dysfunctional uterine bleeding and endometriosis, and as palliative treatment of endometrial cancer, renal cell carcinoma, breast cancer, and prostate cancer. High-dose acetate is used to treat anorexia, cachexia, and AIDS related wasting.

CHAPTER 7

Enzymatic browning: a chemical process, involving polyphenol oxidase, catechol oxidase and other enzymes that create melanins and benzoquinone from natural phenols, resulting in a brown color. Enzymatic browning generally requires exposure to oxygen, thus the browning that occurs when an apple, for example, is cut.

Ewe: a female sheep, especially when full grown.

Micronutrients: nutrients required by humans and other organisms throughout life in small quantities to orchestrate a range of physiological functions. For people, they include dietary trace minerals in amounts generally less than 100 milligrams/day as opposed to macro minerals which are required in larger quantities. The micro minerals or trace elements include at least iron, cobalt, chromium, copper, iodine, manganese, selenium, zinc and molybdenum. Micronutrients also include vitamins, which are organic compounds required as nutrients in tiny amounts by an organism.

Neonate: a newborn infant, especially one less than four weeks old.

Syndesmochorial: relating to the placenta in ruminant animals.

CHAPTER 8

Aneurysm or **aneurism**: a localized, blood-filled balloon-like bulge in the wall of a blood vessel.

***Autosome*:** a chromosome that is not an allosome (i.e., not a sex chromosome). Autosomes appear in pairs whose members have the same form but differ from other pairs in a diploid cell, whereas members of an allosome pair may differ from one another and thereby determine sex. For example, humans have a diploid genome that usually contains 22 pairs of autosomes and one allosome pair (46 chromosomes total). The autosome pairs are labeled with numbers (1-22 in humans) roughly in order of their sizes in base pairs, while allosome are labeled with their letters. By contrast, the allosome pair consists of two X chromosomes in females or one X and one Y chromosome in males. (Unusual combinations of XYY, XXY, XXX, XXXX, XXXXX or XXYY, among other allosomecombinations, are known to occur and usually cause developmental abnormalities.)

Corpus luteum: (Latin for "yellow body") (plural corpora lutea) a temporary endocrine structure in female mammals that is involved in the production of relatively high levels of progesterone and moderate levels of estradiol and inhibin A. It is colored as a result of concentrating carotenoids from the diet and secretes a moderate amount of estrogen to inhibit further release of Gonadotropin releasing hormone (GnRH) and thus secretion of Luteinizing hormone (LH) and Follicle stimulating hormone (FSH).

Cytochrome c oxidase or Complex ***IV***: EC 1.9.3.1) a large transmembrane protein complex found in bacteria and the mitochondrion.

Embryo: a multicellular diploid eukaryote in its earliest stage of development, from the time of first cell division until birth, hatching, or germination. In humans, it is called an embryo until about eight weeks after fertilization (i.e. ten weeks after the last menstrual period or LMP), and from then it is instead called a fetus.

Enzymes: large biological molecules responsible for the thousands of chemical interconversions that sustain life. They are highly selective catalysts, greatly accelerating both the rate and specificity of metabolic reactions, from the digestion of food to the synthesis of DNA. Most enzymes are proteins, although some catalytic RNA molecules have been identified. Enzymes adopt a specific three dimensional structure, and may employ organic (e.g. biotin) and inorganic (e.g. magnesium ion) cofactors to assist in catalysis.

Estradiol: the most potent estrogen in humans; pharmacologically, it is often used in the form of its esters or as a semisynthetic derivative *(ethinyl e.)* For properties and uses, see estrogen.

Estrogens: or **Oestrogens** a group of compounds named for their importance in both menstrual and estrous reproductive cycles. They are the primary female sex hormones. Natural estrogens are steroid hormones, while some synthetic ones are non-steroidal. The name comes from the Greek (oistros), literally meaning "gadfly" but figuratively sexual passion or desire and the suffix-gen, meaning "producer of".

Gamete: (from Ancient Greek *gametes* "husband" *gamete* "wife") a cell that fuses with another cell during fertilization (conception) in organisms that reproduce sexually. In species that produce two morphologically distinct types of gametes, and in which each individual produces only one type, a female is any individual that produces the larger type of gamete called an ovum (or egg) and a male produces the smaller tadpole like type called a sperm. This is an example of anisogamy or heterogamy, the condition in where females and males produce gametes of different sizes (this is the case in humans; the human ovum has approximately 100,000 times the volume of a single human sperm cell). In contrast, isogamy is the state of gametes from both sexes being the same size and shape, and given arbitrary designators for mating type. The name gamete was introduced by the Austrian biologist Gregor Mendel. Gametes carry half the genetic information of an individual, 1 in of each type.

Gland: an organ in an animal's body that synthesizes a substance for release of substances such as hormones or breast milk, often into the bloodstream (endocrine gland) or into cavities inside the body or its outer surface (exocrine gland).

Granulosa cell or **follicular cell:** a somatic cell of the sex cord that is closely associated with the developing female gamete (called an oocyte or egg) in the ovary of mammals.

Hormone: a chemical released by a cell, a gland, or an organ in one part of the body that affects cells in other parts of the organism. Generally, only a small amount of hormone is required to alter cell metabolism. In essence, it is a chemical messenger that transports a signal from one cell to another. All multicellular organisms produce hormones; plant hormones are also called phytohormones.

Leydig cells: also known as interstitial cells of Leydig, are found adjacent to the seminiferous tubules in the testicle. They produce testosterone in the presence of luteinizing hormone (LH). Leydig cells are polyhedral in shape, display a large prominent nucleus, an eosinophilic cytoplasm and numerous lipid-filled vesicles.

Lysyl oxidase: also known as **protein**-lysine 6- oxidase, is a protein that, in humans, is encoded by the *LOX* gene. Its inhibition can cause lathyrism, but, at the same time, its upregulation by tumor cells may promote metastasis of the existing tumor, causing it to become malignant and cancerous.

Organism: any contiguous living system (such as animal, fungus, micro-organism, or plant). In at least some form, all types of organisms are capable of responding to stimuli, reproduction, growth and development, and maintenance of homeostasis as a stable whole.

Progesterone: also known as P4 (pregn-4-ene-3,20- dione) a C-21 steroid hormone involved in the female menstrual cycle, pregnancy (supports gestation) and embryogenesis of humans and other species. Progesterone belongs to a class of hormones called progestogens, and is the major naturally occurring human progestogen.

Testosterone: a steroid hormone from the androgen group and is found in mammals, reptiles, birds, and other vertebrates. In mammals, testosterone is secreted primarily in the testicles of males and the ovaries of females, although small amounts are also secreted by the adrenal glands. It is the principal male sex hormone and an anabolic steroid.

Thrombosis: the formation of a blood clot inside a blood vessel, obstructing the flow of blood through the circulatory system. When a blood vessel is injured, the body uses platelets (thrombocytes) and fibrin to form a blood clot to prevent blood loss.

Sertoli cells: elongated cells in the seminiferous tubules that ensheathe spermatogenic cells, providing a microenvironment that supports spermiogenesis and spermatocytogenesis; they secrete androgen-binding protein and establish the blood-testis barrier by forming tight junctions with adjacent Sertoli cells.

Superoxide dismutase: enzymes that catalyze the dismutation of superoxide ($O2^{-}$) into oxygen and hydrogen peroxide. Thus, they are an important antioxidant defense in nearly all cells exposed to oxygen. One of the exceedingly rare exceptions is Lactobacillus plantarum and related lactobacilli, which use a different mechanism.

Follicle: (anatomy), a small spherical group of cells containing activity.

CHAPTER 9

Androgen: a steroid hormone, such as testosterone or androsterone that controls the development and maintenance of masculine characteristics. Also called *androgenic hormone*.

Antiferromagnetism*: *General Physic: the phenomenon exhibited by substances that resemble paramagnetic substances in the value of their relative permeability but that behave like ferromagnetic substances when their temperature is varied.

Diamagnetism: the property of being repelled by both poles of a magnet. Most substances commonly considered to be nonmagnetic, such as water, are actually diamagnetic. Though diamagnetism is a very weak effect compared with ferromagnetism and paramagnetism, it can be used to levitate objects.

Dynamo theory: geomagnetic dynamo in geophysics, dynamo theory proposes a mechanism by which a celestial body such as the Earth or a star generates a magnetic field. The theory describes the process through which a rotating, convecting, and electrically conducting fluid can maintain a magnetic field over astronomical time scales.

Ferromagnetism: the property of being strongly attracted to either pole of a magnet. Ferromagnetic materials, such as iron, contain unpaired electrons, each with a small magnetic field of its own, that align readily with each other in response to an external magnetic field. This alignment tends to persists even after the magnetic field is removed, a phenomenon called hysteresis. Ferromagnetism is important in the design of electromagnets, transformers, and many other electrical and mechanical devices, and in analyzing the history of the earth's magnetic reversals.

Gamma radiation: also known as gamma rays, and denoted by the Greek letter γ, refers to electromagnetic radiation of high frequency and therefore high energy per photon. Gamma rays are ionizing radiation, and are thus biologically hazardous. They are classically produced by the decay from high energy states of atomic nuclei (gamma decay), but are also created by other processes. Paul

Villard, a French chemist and physicist, discovered gamma radiation in 1900, while studying radiation emitted from radium.

Hydrogen: *Symbol (*H) a colorless, highly flammable gaseous element, the lightest of all gases and the most abundant element in the universe, used in the production of synthetic ammonia and methanol, in petroleum refining, in the hydrogenation of organic materials, as a reducing atmosphere, in oxyhydrogen torches, and in rocket fuels.

Labium: any of four folds of tissue of the female external genitalia.

Magnetic* field: *General Physics: a field of force surrounding a permanent magnet or a moving charged particle, in which another permanent magnet or moving charge experiences a force.

Magnetism: ***General Physics***: the property of attraction displayed by magnets.

Mammary *gland*: any of the milk-producing glands in female mammals, consisting of lobes containing clusters of alveoli with a system of ducts to convey the milk to an external nipple or teat. These glands typically occur in pairs and begin secreting milk when young are born.

Menarche: the first menstrual period, usually occurring during puberty.

Paramagnetism: general physics the phenomenon exhibited by substances that have a relative permeability slightly greater than unity and a positive susceptibility. The effect is the alignment of unpaired spins of electrons in atoms of the material.

Uterus: a hollow muscular organ located in the pelvic cavity of female mammals in which the fertilized egg implants and develops. Also called *womb*.

CHAPTER 10

Androgen: a steroid hormone, such as testosterone or androsterone that controls the development and maintenance of masculine characteristics. Also called *androgenic hormone*.

Anterior: placed before or in front. Located on or near the front of an organ or on the ventral surface of the body in humans

Estrogen: any of several steroid hormones produced chiefly by the ovaries and responsible for promoting estrus and the development and maintenance of female secondary sex characteristics. It is also produced synthetically for use in medical treatment.

Gynecomastia: abnormal enlargement of the breasts in a male.

Luteinizing hormone: a glycoprotein hormone secreted by the adenohypophysis of vertebrates which stimulates hormone production by interstitial cells of gonads. Abbreviated LH. Also known as interstitial-cell-stimulating hormone (ICSH).

Testosterone: a white crystalline steroid hormone produced primarily in the testes and responsible for the development and maintenance of male secondary sex.

Seminiferous *tubule*: one of two or three twisted, curved tubules in each lobule of the testis in which spermatozoa develop.

Anemia: a pathological deficiency in the oxygen- carrying component of the blood, measured in unit volume concentrations of hemoglobin, red blood cell volume, or red blood cell number.

Fibrocystic: characterized by increased fibrosis and cystic spaces, especially in glandular tissue: *fibrocystic disease of the breast.*

Neuralgia: sharp, severe paroxysmal pain extending along a nerve or group of nerves. Peptide secreted by the hypothalamus which acts primarily to inhibit the release of growth hormone from the anterior pituitary.

Palsy: complete or partial muscle paralysis, often accompanied by loss of sensation and uncontrollable body movements or tremors.

Parathyroid: of, relating to, or obtained from the parathyroid glands. A parathyroid hormone.

Presenile: pertaining to a condition resembling senility, but occurring in early or middle life.

Somatostatin: a polypeptide hormone produced chiefly by the hypothalamus that inhibits the secretion of various other hormones, such as somatotropin, glucagon, insulin, thyrotropin, and gastrin.

CHAPTER 11

Anemia: a pathological deficiency in the oxygen- carrying component of the blood, measured in unit volume concentrations of hemoglobin, red blood cell volume, or red blood cell number.

Fibrocystic: characterized by increased fibrosis and cystic spaces, especially in glandular tissue: *fibrocystic disease of the breast.*

Neuralgia: sharp, severe paroxysmal pain extending along a nerve or group of nerves. Peptide secreted by the hypothalamus which acts primarily to inhibit the release of growth hormone from the anterior pituitary.

Palsy: complete or partial muscle paralysis, often accompanied by loss of sensation and uncontrollable body movements or tremors.

Parathyroid: of, relating to, or obtained from the parathyroid glands. A parathyroid hormone.

Presenile: pertaining to a condition resembling senility, but occurring in early or middle life.

Somatostatin: a polypeptide hormone produced chiefly by the hypothalamus that inhibits the secretion of various other hormones, such as somatotropin, glucagon, insulin, thyrotropin, and gastrin.

CHAPTER 12

Acidosis: the accumulation of acid and hydrogen ions or depletion of the alkaline reserve (bicarbonate content) in the blood and body tissues, decreasing the ph.

Carbon dioxide: an odorless, colorless gas, CO2, resulting from oxidation of carbon, and formed in the tissues and eliminated by the lungs; used in some pump oxygenators to maintain blood carbon dioxide tension.

Carbohydrate: any of a group of organic compounds that includes sugars, starches, celluloses, and gums and serves as a major energy source in the diet of animals; they are produced by photosynthetic plants and contain only carbon, hydrogen, and oxygen, usually in the ratio 1:2:1.

Bile: a fluid secreted by the liver, concentrated in the gallbladder, and poured into the small intestine via the bile ducts, which helps in alkalinizing the intestinal contents and plays a role in emulsification, absorption, and digestion of fat; its chief constituents are conjugated bile salts, cholesterol, phospholipid, bilirubin, and electrolytes.

Carbon monoxide: an odorless gas, CO, formed by burning carbon or organic fuels with a scanty supply of oxygen; inhalation causes central nervous system damage and asphyxiation by combining irreversibly with blood hemoglobin.

Carbon *tetrachloride*: a clear, colorless, volatile liquid; inhalation of its vapors can depress central nervous system activity and cause degeneration of the liver and kidneys.

Glucose: a six-carbon aldose occurring as the d- form and found as a free monosaccharide in fruits and other plants or combined in glucosides and polysaccharides. It is the end product of carbohydrate metabolism, and is the chief source of energy for living organisms, its utilization being controlled by insulin. Excess glucose is converted to glycogen and stored in the liver and muscles for use as needed and, beyond that, is converted to fat and stored as adipose tissue. Glucose appears in the urine in diabetes mellitus. In pharmaceuticals, called dextrose.

Proteins: macromolecules made of up of amino acids joined by peptide linkages and that provide essential life functions. Major sources of dietary proteins include meat, fowl, fish, eggs, beans, seeds, nuts, legumes, and dairy products.

CHAPTER 13

Arginase: a manganese containing enzyme. The reaction catalyzed by this enzyme is: arginine + $\underline{H2O}$ → ornithine + urea. It is the final enzyme of the urea cycle.

Amino acids: biologically important organic compounds made from amine (-NH_2) and carboxylic acid (-COOH) functional groups, along with a side chain specific to each amino acid. The key elements of an amino acid are carbon, hydrogen, oxygen, and nitrogen, though other elements are found in the side-chains of certain amino acids. About 500 amino acids are known and can be classified in many ways. Structurally they can be classified according to the functional groups' locations as alpha- (α-), beta- (β-), gamma- (γ-) or delta- (δ-) amino acids; other categories relate to polarity, pH level, and side chain group type (aliphatic, acyclic, aromatic, containing hydroxyl or sulfur, etc.) In the form of proteins, amino acids comprise the second largest component (after water) of human muscles, cells and other tissues. Outside proteins, amino acids perform critical roles in processes such as neurotransmitter transport and biosynthesis.

Aromatic: ***Chemistry***: (of an organic compound) having an unsaturated ring containing alternating double and single bonds, esp. containing a benzene ring; exhibiting aromaticity.

Bacteria: any of numerous groups of microscopic one-celled organisms constituting the phylum Schizomycota, of the kingdom Monera, various species of which are involved in infectious diseases, nitrogen fixation, fermentation, or putrefaction.

Carboxylation: in chemistry is a chemical reaction in which a carboxylic acid group is introduced in a substrate. The opposite reaction is decarboxylation.

Cyclohexane: a cycloalkane with the molecular formula C_6H_{12}. Cyclohexane is used as a nonpolar solvent for the chemical industry, and also as a raw material for the industrial production of adipic acid and caprolactam, both of which being intermediates used in the production of nylon. On an industrial scale, cyclohexane is produced by reacting benzene with hydrogen. Because of its unique chemical and conformational properties, cyclohexane is also used in labs in analysis and as a standard. Cyclohexane has a distinctive detergent- like odor, reminiscent of cleaning product (in which it is sometimes used).

Hydrogen: a chemical element with chemical symbol (H) and atomic number1. Hydrogen is the lightest element and its monatomic form (H1) is the most abundant chemical substance, constituting roughly 75% of the Universe's baryonic mass. Non- remnant stars are mainly composed of hydrogen in its plasma state.

Macromolecule: a very large molecule commonly created by polymerization of smaller subunits. In biochemistry, the term is applied to the three conventional biopolymers (nucleic acids, proteins and carbohydrates), as well as non-polymeric molecules with large molecular mass such as lipids and macrocycles. The individual constituent molecules of macromolecules are called monomers (*mono=single, meros=part*).

Meconium: a greenish mass of mucous, desquamated epithelial cells, lanugo, and vernix caseosa that collects in the fetal intestine, becoming the first fecal discharge of the newborn.

Phytin: the calcium magnesium salt of phytic acid; a widely distributed natural compound found in plants. Large quantities of Phytin are found in bran and in oil cakes of cotton, hemp, and other plants. A therapeutic agent called Phytin is obtained from the substances. A white, amorphous, odorless powder poorly soluble in water, it is used in medicine to stimulate hematopoiesis, accelerate bone growth and development, and treat some neurological disorders.

Porphyrias: a group of rare inherited or acquired disorders of certain enzymes that normally participate in the production of porphyrins and heme. They manifest with either neurological complications or skin problems or occasionally both.

CHAPTER 14

Aneurysm: a localized, blood-filled dilation of a blood vessel caused by disease or weakening of the vessel wall.

Anticarcinogenic: to produce pertaining to a substance or device that neutralizes the effects of a cancer-causing substance.

Epithelioma: a benign or malignant tumor derived from epithelium.

Etiology: the science and study of the causes or origins of disease. The cause or origin of a disease or disorder as determined by medical diagnosis.

Hemorrhagic colitis: abdominal cramps and bloody diarrhea, without fever, attributed to a self-limited infection by a strain of *Escherichia coli.*

Hydrogen: a flammable colourless gas that is the lightest and most abundant element in the universe. It occurs mainly in water and in most organic compounds and is used in the production of ammonia and other chemicals, in the hydrogenation of fats and oils, and in welding.

Hydrogenous: of, pertaining to, or containing hydrogen.

Lecithin: Is a generic term to designate any group of yellow-brownish fatty substances occurring in animal and plant tissues composed of phosphoric acid, choline, fatty acids, glycerol, glycolipids, triglycerides, and phospholipids (e.g., phosphatidylcholine, phosphatidylethanolamine, and phosphatidylinositol.

Listeria: a genus of bacteria that contains ten species. Named after the English pioneer of sterile surgery Joseph Lister, the genus received its current name in 1940. *Listeria* species are Gram positive bacilli. The major human pathogen in the *Listeria* genus is *L.* monocytogenes. It is usually the causative agent of the relatively rare bacterial disease, listeriosis, a serious infection caused by eating food contaminated with the bacteria. The disease affects primarily pregnant women, newborns, adults with weakened immune systems, and the elderly.

Malachite: light to dark green carbonate mineral, used as a source of copper and for ornamental stoneware.

Myocardial: pertaining to the muscular tissue of the heart (the myocardium). Pertaining to the muscular tissue of the heart.

Myocardial infarction: death of the cells of an area of the heart muscle (MYOCARDIUM) as a result of oxygen deprivation, which in turn is caused by obstruction of the blood supply; commonly referred to as a "heart attack."

Pharmacology: the science of drugs, including their composition, uses, and effects. The characteristics or properties of a drug, especially those that make it medically effective.

Trachoma: (Greek: 'roughness') an infectious disease caused by the Chlamydia trachomatis bacterium which produces a characteristic roughening of the inner surface of the eyelids. Also called granular conjunctivitis and Egyptian ophthalmia, it is the leading cause of infectious blindness in the world. Globally, about 40 million people have an active infection and as many as 8 million people are visually impaired as a result of this disease. It belongs to a group of diseases known as neglected tropical diseases.

Verdigris: the common name for a green pigment obtained through the application of acetic acid to copper plates or the natural patina formed when copper, brass or bronze is weathered and exposed to air or seawater over a period of time. It is usually a basic copper carbonate, but near the sea will be a basic copper chloride. If acetic acid is present at the time of weathering, it may consist of copper (II) acetate.

CHAPTER 15

Algae: any of various chiefly aquatic, eukaryotic, photosynthetic organisms, ranging in size from single-celled forms to the giant kelp.

Chemical: pertaining to chemistry. A substance composed of chemical elements, or obtained by chemical processes.

Chlorophyll: *a*ny of a group of green pigments that are found in the chloroplasts of plants and in other photosynthetic organisms such as cyanobacteria, especially: A waxy blue-black microcrystalline green-plant pigment, C55H72MgN4O5, with a characteristic blue-green alcohol solution. Also called *chlorophyll a*. A similar green-plant pigment, C55H70MgN4O6, having a brilliant green alcohol solution. Also called *chlorophyll b*.

Consciousness: the state or condition of being conscious. A sense of one's personal or collective identity, especially the complex of attitudes, beliefs, and sensitivities held by or considered characteristic of an individual or a group.

Cortisone: pharmacology a glucocorticoid hormone, the synthetic form of which has been used in treating rheumatoid arthritis, allergic and skin diseases, leukemia, etc.; 17-hydroxy-11- dehydrocorticosterone.

Cyanobacteria: blue-green bacteria, unicellular or filamentous organisms that fix both carbon dioxide (in the presence of light) and nitrogen. Several species are common causes of water pollution and cause cyanobacteria poisoning. Formerly called blue green algae.

Deciduous: falling off or shed at a specific stage of growth, as teeth of the first dentition.

7-Dehydrocholesterol: is a zoosterol that functions in the serum as a cholesterol precursor, and is converted to vitamin D3 in the skin, therefore functioning as provitamin-D3. The presence of this compound in human skin enables humans to manufacture vitamin D3 from ultra violet rays in the sun light, via an intermediate isomer pre- vitamin D3. It is also found in the milk of several mammalian species. In insects it is a precursor for the hormone ecdysone, required for reaching adulthood. It was discovered by Nobel-laureate organic chemist Adolf Windaus.

Dermis: is a layer of skin between the epidermis (with which it makes up the cutis) and subcutaneous tissues, that consists of connective tissue and cushions the body from stress and strain. It is divided into two layers, the superficial area adjacent to the epidermis called the papillary region and a deep thicker area known as the reticular dermis. The dermis is tightly connected to the epidermis through a basement membrane.

Structural components of the dermis are collagen, elastic fibers, and extrafibrillar matrix. It also contains Mechanoreceptors that provide the sense of touch and heat, hair follicles, sweat glands, sebaceous glands, apocrine glands, lymphatic vessels and blood vessels. Those blood vessels provide nourishment and waste removal for both dermal and epidermal cells.

Epidermis: the outermost, nonvascular, no sensitive layer of the skin, covering the dermis.

Glycerol: a syrupy, sweet, colorless or yellowish liquid, C3H8O3, obtained from fats and oils as a byproduct of saponification and used as a solvent, an antifreeze, a plasticizer, and a sweetener and in the manufacture of dynamite, cosmetics, liquid soaps, inks, and lubricants.

Lipid: any of a group of organic compounds, including the fats, oils, waxes, sterols, and triglycerides, that are insoluble in water but soluble in nonpolar organic solvents, are oily to the touch, and together with carbohydrates and proteins constitute the principal structural material of living cells.

Melanocyte: an epidermal cell capable of synthesizing melanin. *Anatomy zoology* a cell, usually in the epidermis, that contains melanin.

Metabolism: the chemical processes occurring within a living cell or organism that are necessary for the maintenance of life. In metabolism some substances are broken down to yield energy for vital processes while other substances, necessary for life, are synthesized.

Melanoma: a dark-pigmented benign or malignant tumor that arises from a melanocyte and occurs most commonly in the skin. Malignant melanoma metastasizes quickly and is associated with sun exposure.

Phospholipid: any of various phosphorus- containing lipids, such as lecithin and cephalin that are composed mainly of fatty acids, a phosphate group, and a simple organic molecule. Also called *phosphatide*. Any of a group of compounds composed of fatty acids, phosphoric acid, and a nitrogenous base: important constituents of all membranes.

Phycobilin: any of a class of red or blue-green pigments found in the red algae and cyanobacteria.

Psychological: pertaining to the mind, its mental processes, and its emotional makeup.

Radioactivity: spontaneous emission of radiation, either directly from unstable atomic nuclei or as a consequence of a nuclear reaction. The radiation, including alpha particles, nucleons, electrons, and gamma rays, emitted by a radioactive substance.

Radium: *Symbol (***Ra)** a luminescent, highly radioactive metallic element found in minute amounts in uranium ores, used as a neutron source for some research purposes, and formerly used in cancer radiotherapy; its most stable isotope is Ra 226 with a half-life of 1,622 years. Atomic number 88.

Serum: the clear yellowish fluid obtained upon separating whole blood into its solid and liquid components after it has been allowed to clot. Also called *blood serum*. Blood serum from the tissues of immunized animals, containing antibodies and used to transfer immunity to another individual.

Steroid: any of numerous naturally occurring or synthetic fat-soluble organic compounds having as a basis 17 carbon atoms arranged in four rings and including the sterols and bile acids, adrenal and sex hormones, certain natural drugs such as digitalis compounds, and the precursors of certain vitamins.

Stoma: a miniature opening in leaves, stems, etc., through which gases are exchanged.

Synthesis: the combining of separate elements or substances to form a coherent whole.

Triglycerides: fatty compounds synthesized from carbohydrates during the process of digestion and stored in the body's adipose (fat) tissues. High levels of triglycerides in the blood are associated with insulin resistance.

Ultraviolet: *abbr.* (**UV)** of or relating to the range of invisible radiation wavelengths from about 4 nanometers, on the border of the x-ray region, to about 380 nanometers, just beyond the violet in the visible spectrum. Ultraviolet light or the ultraviolet part of the spectrum.

Velocity: rapidity or speed of motion; specifically, the distance traveled per unit time.

Wavelength: the distance between a given point on one wave cycle and the corresponding point on next successive wave cycle. A pure color is produced by light of a specific wavelength. Electromagnetic waves of different wavelengths account for many of the transmission characteristics of radio and television.

Xylem: ***Botany***: a plant tissue that conducts water and mineral salts from the roots to all other parts, provides mechanical support, and forms the wood of trees and shrubs. It is of two types (protoxylem and metaxylem), both of which are made up mainly of vessels and tracheids. See also protoxylem, metaxylem.

CHAPTER 16

Galaxy: any of numerous large-scale aggregates of stars, gas, and dust that constitute the universe, containing an average of 100 billion (10^{11}) solar masses and ranging in diameter from 1,500 to 300,000 light-years. Also called *nebula*'.

Photosynthesis: the process by which green plants, algae, diatoms, and certain forms of bacteria make carbohydrates from carbon dioxide and water in the presence of chlorophyll, using energy captured from sunlight by chlorophyll, and releasing excess oxygen as a byproduct. In plants and algae, photosynthesis takes place in organelles called chloroplasts. Photosynthesis is usually viewed as a two-step process. First, in the light reactions, the energy-providing molecule ATP is synthesized using light energy absorbed by chlorophyll and accessory pigments such as carotenoids and phycobilins, and water is broken apart into oxygen and a hydrogen ion, with the electron of the hydrogen transferred to another energy molecule, NADPH. The ATP and NADPH molecules power the second part of photosynthesis by the transfer of electrons. In these light-independent or dark reactions, carbon is broken away from carbon dioxide and combined with hydrogen via the Calvin cycle to create carbohydrates. Some of the carbohydrates, the sugars, can then be transported around the organism for immediate use; others, the starches, can be stored for later use.

Vegetation: the act or process of vegetating.

CHAPTER 17

Adenosine: a nucleoside, C10H13N5O4, composed of adenine linked to ribose that is a structural component of nucleic acids and the major molecular component of ADP, AMP, and ATP.

Atom: a part or particle considered to be an irreducible constituent of a specified system. A unit of matter, the smallest unit of an element, having all the characteristics of that element and consisting of a dense, central, positively charged nucleus surrounded by a system of electrons. The entire structure has an approximate diameter of 10^{-8} centimeter and characteristically remains undivided in chemical reactions except for limited removal, transfer, or exchange of certain electrons.

Cell: the basic unit of living matter in all organisms, consisting of protoplasm enclosed within a cell membrane. All cells except bacterial cells have a distinct nucleus that contains the cell's DNA as well as other structures (called organelles) that include mitochondria, the endoplasmic reticulum, and vacuoles. The main source of energy for all of a cell's biological processes is ATP. Any of various devices, or units within such devices, that are capable of converting some form of energy into electricity. Cells contain two electrodes and an electrolyte.

Chromosome: a threadlike linear strand of DNA and associated proteins in the nucleus of eukaryotic cells that carries the genes and functions in the transmission of hereditary information. A circular strand of DNA in bacteria that contains the hereditary information necessary for cell life.

Chromosphere: an incandescent, transparent layer of gas, primarily hydrogen, several thousand miles in depth, lying above and surrounding the photosphere of a star, such as the sun, but distinctly separate from the corona.

DNA: Deoxyribonucleic acid: an extremely long, double-stranded nucleic acid molecule arranged as a double helix that is the main constituent of the chromosome and that carries the genes as segments along its strands found chiefly in the chromatin of cells and in many viruses.

Ellipse: a plane curve, especially: **a.** a conic section whose plane is not parallel to the axis, base, or generatrix of the intersected cone.**b.** The locus of points for which the sum of the distances from each point to two fixed points is equal.

Frequency: the property or condition of occurring at frequent intervals.

Glucose: a monosaccharide sugar, C6H12O6, occurring widely in most plant and animal tissue. It is the principal circulating sugar in the blood and the major energy source of the body.

Helix: ***Mathematics***: a three-dimensional curve that lies on a cylinder or cone, so that its angle to a plane perpendicular to the axis is constant.**2.** A spiral form or structure.

Hydrogen: *Symbol (***H**) a colorless, highly flammable gaseous element, the lightest of all gases and the most abundant element in the universe, used in the production of synthetic ammonia and methanol, in petroleum refining, in the hydrogenation of organic materials, as a reducing atmosphere, in oxyhydrogen torches, and in rocket fuels. Atomic number 1.

Membrane: a thin, pliable layer of tissue covering surfaces or separating or connecting regions, structures, or organs of an animal or a plant.

Mitochondrion: a spherical or elongated organelle in the cytoplasm of nearly all eukaryotic cells, containing genetic material and many enzymes important for cell metabolism, including those responsible for the conversion of food to usable energy. Also called *chondriosome*.

Monomer: a molecule that can combine with others to form a polymer.

Nucleus: a central or essential part around which other parts are gathered or grouped; a core: A large, membrane-bound, usually spherical protoplasmic structure within a living cell, containing the cell's hereditary material and controlling its metabolism, growth, and reproduction.

Photon: the quantum of electromagnetic energy, regarded as a discrete particle having zero mass, no electric charge, and an indefinitely long lifetime. A unit of retinal illumination, equal to the amount of light that reaches the retina through 1 square millimeter of pupil area from a surface having a brightness of 1 candela per square meter.

Phosphate: any salt or ester of any phosphoric acid, esp. a salt of orthophosphoric acid. Any of several chemical fertilizers containing phosphorous compounds.

Phyllotaxis: the arrangement of leaves on a stem.

Polarity: intrinsic polar separation, alignment, or orientation, especially of a physical property: *magnetic polarity; ionic polarity.* The possession or manifestation of two opposing attributes, tendencies, or principles.

Polymer: any of numerous natural and synthetic compounds of usually high molecular weight consisting of up to millions of repeated linked units, each a relatively light and simple molecule.

CHAPTER 18

Acetylcholine: is an organic molecule that acts as a neurotransmitter in many organisms, including humans. It is an ester of acetic acid and choline. Parts of the body that use or are affected by acetylcholine are referred to as cholinergic activity are called anticholinergics.

Adenosine: plays an important role in biochemical processes, such as energy transfer as adenosine triphosphate (ATP) and adenosine diphosphate (ADP) as well as in signal transduction as cyclic adenosine monophosphate. It is also an inhibitory neurotransmitter, believed to play a role in promoting sleep and suppressing arousal.

Amplitude: largeness, fullness; wideness in range or extent. In a phenomenon that occurs in Waves, the maximal deviation of a wave from the baseline.

Basal ganglia: the thalamus together with other closely related masses of grey matter, situated near the base of the brain hard, shell-like covering, such as that enclosing the kernel of a peach.

Blood-brain barrier: a blockade of cells separating the circulating blood from elements of the central nervous system (CNS); it acts as a filter, preventing many substances from entering the central nervous system.

Caudate nucleus: a basal ganglion located in the lateral ventricle of the brain that has a curved, tail like extension and functions in motor control.

Cerebellum: the trilobed structure of the brain, lying posterior to the pons and medulla oblongata and inferior to the occipital lobes of the cerebral hemispheres, that is responsible for the regulation and coordination of complex voluntary muscular movement as well as the maintenance of posture and balance.

Circadian rhythm: a daily rhythmic activity cycle, based on 24-hour intervals, that is exhibited by many organisms.

Coccyx: plural coccyges, commonly referred to as the tailbone, is the final segment of the vertebral column in tailless primates. Comprising three to five separate or fused vertebrae (the coccygeal vertebrae) below the sacrum, it is attached to the sacrum by a fibrocartilaginous joint, the sacrococcygeal symphysis, which permits limited movement between the sacrum and the coccyx.

Diencephalon: the posterior part of the forebrain, consisting of the hypothalamus, thalamus, metathalamus, and epithalamus; the subthalamus is often recognized as a distinct division. Diencephalon, Posterior (dorsal) *(A)* and anterior (inferior) *(B)* views of the base of the brain, showing the diencephalon in relation to the mesencephalon (midbrain) and rhombencephalon (hindbrain). The posterior of the two brain vesicles formed by specialization in embryonic development.

Dopamine: a monoamine neurotransmitter formed in the brain by the decarboxylation of Dopa and essential to the normal functioning of the central nervous system. A reduction in its concentration within the brain is associated with Parkinson's disease.

Dorsal: (From Latin *dorsum* 'back') (anatomy), the back of an animal or human being.

Entrainment Biomusicology: the synchronization of organisms to an external rhythm.

Entrainment Chronobiology: the alignment of a circadian system's period and phase to the period and phase of an external rhythm.

Erebral: of or relating to the brain or cerebrum. Appealing to or requiring the use of the intellect; intellectual rather than emotional.

Estrogen: any of a class of hormones that primarily influence the female reproductive system's development, maturation, and function. The three major estrogens, estradiol, estrone, and estriol are produced mainly by the ovaries and placenta; the adrenal glands and the testes secrete smaller amounts. Estrogens affect the ovaries, vagina, fallopian tubes, uterus, and mammary glands and play crucial roles in puberty, menstruation, pregnancy, and parturition (labour). They also influence the structural differences between female and male bodies. In experimental animals, loss of estrogens diminishes mating desires and other behavioral patterns.

Forebrain: the most anterior of the three primary regions of the embryonic brain from which the telencephalon and diencephalon develop. The segment of the adult brain that develops from the embryonic forebrain and includes the cerebrum, thalamus, and hypothalamus.

Gamma amino butyric acid: *Abbr.* **GABA** an amino acid that occurs in the central nervous system and that is associated with transmission of inhibitory nerve impulses.

Ganglion: an encapsulated collection of nerve-cell bodies, usually located outside the brain and spinal cord.

Globus pallidus: the inner pale yellow part of the lenticular nucleus.

Grey matter: the greyish tissue of the brain and spinal cord, containing nerve cell bodies, dendrites, and bare (unmyelinated) axons. Technical name: substantia grisea Compare white matter.

Kundalini: energy that lies dormant at the base of the spine until it is activated, as by the practice of yoga, and channeled upward through the chakras in the process of spiritual perfection.

Locus Coeruleus: (also spelled locus caeruleus) is a nucleus in the pons (part of the brainstem) involved with physiological responses to stress and panic. Is the principal site for brain synthesis of norepinephrine (noradrenaline). The locus Coeruleus and the areas of the body affected by the norepinephrine it produces are described collectively as the locus Coeruleus-noradrenergic system or LC-NA system. Norepinephrine may also be released directly into the blood from the adrenal medulla.

Hindbrain: the portion of the embryonic brain from which the metencephalon and myelencephalon develop. The lower or hind region of the adult brain comprising the pons and medulla oblongata.

Inferior colliulus: the ovoid, paired, inferior eminence of the laminae of mesencephalic tectum; it receives the lateral lemniscus and projects by way of the brachium of inferior colliculus to the medial geniculate body of the thalamus, and is thus an essential way station in the central auditory pathway.

Melanin: any of a group of naturally occurring dark pigments, especially the pigment found in skin, hair, fur, and feathers.

Melanoma: pathology a malignant tumor composed of melanocytes, occurring esp. in the skin, often as a result of excessive exposure to sunlight.

Melanopsin: is a photopigment found in specialized photosensitive ganglion cells of the retina that are involved in the regulation of circadian rhythms, pupillary light reflex, and other non-visual responses to light. In structure, melanopsin is an opsin, a retinylidene protein variety of G-protein- coupled receptor. Melanopsin is most sensitive to blue light. A melanopsin based receptor has been linked to the association between light sensitivity and migraine pain.

Melatonin: a catecholamine hormone synthesized and released by the pineal body; in mammals it influences hormone production and in many species regulates seasonal changes such as reproductive pattern and fur color. In humans it is implicated in the regulation of sleep, mood, puberty, and ovarian cycles and it has been tried therapeutically for insomnia, jet lag, and other conditions.

Mesencephalon: the midbrain.

Midbrain: the portion of the vertebrate brain that develops from the middle section of the embryonic brain.

Medulla: refers to the middle of something and derives from the Latin word for marrow. Its anatomical uses include: Medulla, a part of the brain stem Medulla spinal is, an alternative name for the spinal cord.

Modulation: the act of tempering.

Molecule: a small mass of matter the smallest amount of a substance which can exist alone; an Aggregation of atoms, specifically a chemical combination of two or more atoms forming a specific chemical substance molecular.

Neurons: nerve cells in the brain, brain stem, and spinal cord that connect the nervous system and the muscles.

Neurotransmitter: a chemical substance, such as acetylcholine or dopamine, which transmits nerve impulses across a synapse.

Oculomotor nerve: either of the third pair of cranial nerves, which originate in the midbrain and control most of the muscles that move the eyeballs.

Optic chiasm: or optic chiasma (Greek "crossing", from the Greek 'to mark with an X', after the Greek letter 'X', chi) is the part of the brain where the optic nerves (CN II) partially cross. The optic chiasm is located at the bottom of the brain immediately below the hypothalamus.

Parasympathetic nervous system: (PSNS, or occasionally PNS) is one of three main divisions of the autonomic nervous system (ANS), the other two being the sympathetic and enteric systems. The ANS is responsible for regulation of internal organs and glands, which occurs unconsciously.

To be specific, the parasympathetic system is responsible for stimulation of "rest-and-digest" or "feed and breed "activities that occur when the body is at rest, especially after eating, including sexual arousal, salivation, lacrimation (tears), urination, digestion and defecation. Its action is described as being complementary to that of one of the other main branches of the ANS, the sympathetic nervous system, which is responsible for stimulating activities associated with the fight or flight response.

Photoreceptors: specialized nerve cells (rods and cones) in the retina that are responsible for Vision.

Pineal gland: a small, unpaired, flattened body shaped something like a pinecone; attached at its anterior pole to the region of the posterior and habenular commissures, and lying in the depression between the two superior colliculi below the splenium of the corpus callosum. It is a glandular structure, composed of follicles containing epithelioid cells and lime concentrations called 'brain sand'; despite its attachment to the brain, it appears to receive nerve fibers exclusively from the peripheral autonomic nervous system. It produces melanin.

Pituitary adenylate cyclase-activating polypeptide: also known as PACAP is a protein that in humans is encoded by the ADCYAP1 gene. PACAP is similar to vasoactive intestinal peptide. One of its effects is to stimulate enterochromaffin like cells. It binds to vasoactive intestinal peptide receptor and to the PACAP receptor.

Pons: a band of nerve fibers on the ventral surface of the brain stem that links the medulla oblongata and the cerebellum with upper portions of the brain. Also called *pons Varolii*.

Putamen: a the outer reddish part of the lenticular nucleus of the basal ganglion - any of several masses of subcortical grey matter at the base of each cerebral hemisphere that seem to be involved in the regulation of voluntary movement.

Red nucleus: a large, well defined, somewhat elongated cell mass of reddish-gray hue that is located in the mesencephalon tegmentum, receives a massive projection from the contralateral half of the cerebellum, receives an additional projection from the ipsilateral motor cortex, and whose efferent connections are with the contralateral half of the rhombencephalic reticular formation and spinal cord.

Retinohypothalamic tract: the retinohypothalamic tract is a photic input pathway involved in the circadian rhythms of mammals. The origin of the retinohypothalamic tract is the intrinsically photosensitive retinal ganglion cells (ipRGC), which contain the photopigment melanopsin. The axons of the ipRGCs belonging to the retinohypothalamic tract project directly, monosynaptically, to the suprachiasmatic nuclei via the optic nerve and the optic chiasm. The suprachiasmatic nuclei receive and interpret information on environmental light, dark and day length, important in the entrainment of the "body clock". They can coordinate peripheral "clocks" and direct the pineal gland to secrete the hormone melatonin.

Reticular formation: a diffuse network of nerve fibers and cells in parts of the brainstem, important in regulating consciousness or wakefulness.

Retina: the innermost tunic of the eyeball, containing the neural elements for reception and transmission of visual stimuli.

Rubrospinal: pertaining to the red nucleus and the spinal cord.

Schistocyte: a red blood cell undergoing fragmentation.

Serotonin: a hormone and neurotransmitter, 5- hydroxytryptamine (5-HT), found in many tissues, including blood platelets, intestinal mucosa, the pineal body, and the central nervous system; it has many physiologic properties including inhibition of gastric secretion, stimulation of smooth muscles, and production of vasoconstriction.

Soma: the entire body of an organism, exclusive of the germ cells.

Striatum: a striped mass of white and grey matter located in front of the thalamus in each cerebral hemisphere; consists of the caudate nucleus and the *lenticular nucleus.*

Spinal cord: elongated nerve bundles that lie in the vertebral canal and from which the spinal nerves emerge.

Substantia nigra: a layer of large pigmented nerve cells in the midbrain that produce dopamine and whose destruction is associated with Parkinson's disease.

Superior cerebellar peduncle: a large bundle of nerve fibers that originates from the dentate nucleus of the cerebellum and emerges from the cerebellum along the lateral wall of the fourth ventricle.

Superior colliculus: the paired, larger, rounded anterior eminence of the laminae of mesencephalic tectum; major afferent connections of the superficial layers are the retina and striate cortex; input to deep layers of the colliculus are polymodal. Its efferent connections are with the lower brainstem and spinal cord (tectobulbar tract and tectospinal tract) and with the pulvinar and other cell groups in the caudal part of the thalamus; it participates in the extra geniculate visual pathway. Synonym(s): colliculus inferior [TA], corpus quadrigeminum posterius, inferior nasal colliculus, Posterior quadrigeminal body.

***Suprachiasmatic nucleus*:** anatomic nucleus which innervates the pineal gland; thought to play a part in the management of circadian rhythms.

Tegmentum: a part of the midbrain consisting of white fibers running lengthwise through gray matter.

Testosterone: a potent steroid hormone secreted mainly by the testes. It can be extracted from the testes of animals or synthesized and used to treat androgen deficiency or promote anabolism.

Thalamus: a large ovoid mass of gray matter situated in the posterior part of the forebrain that relays sensory impulses to the cerebral cortex, the middle part of the diencephalon of the brain, serving to transmit and integrate sensory impulses.

Triphosphate: a salt or ester containing three phosphate groups.

Trochlear: of, resembling, or situated near a trochlea. Of or relating to the trochlear nerve.

Tryptophan: an essential amino acid, $C_{11}H_{12}N_2O_2$, formed from proteins during digestion by the action of proteolytic enzymes. It is necessary for normal growth and development and is the precursor of several substances, including serotonin and niacin.

Tyrosinase: a copper-containing enzyme of plant and animal tissues that catalyzes the production of melanin and other pigments from tyrosine by oxidation, as in the blackening of a peeled or sliced potato exposed to air. An enzyme occurring in many organisms that is a catalyst in the conversion of tyrosine to the pigment melanin; inactivity of this enzyme results in albinism.

Vegetation: the act or process of vegetating. The plants of an area or a region; plant life. *Pathology* An abnormal growth on the body.

Vertebrate: having a spinal column (vertebrae). An animal with a vertebral column; any member of the Vertebrata.

Definitions are from: *www.freeonlineDictionary.com*

BIBILIOGRAPHY

Wallach, Joel D, and Ma Lan. *Dead Doctors Don't Lie*. Franklin, Tenn. Legacy Communications Group, 1999.

Griaule, Marcel. *Conversations with Ogotemmeli: An Introduction to Dogon Religious Ideas.* London: Oxford University Press 1965

Challoner, Jack. *The Visual Dictionary of Chemistry*. DK Publishing, 1996

Griaule, Marcel, Dieterlen, Germaine. *The Pale Fox.* Afrikan World books 1965

Rubin, Herbert H. *Eugenic and Sex Harmony*. New York: Elliot Publishing Company, 1936

Moritz, Andrea. *The Amazing Liver and Gallbladder Flush*. Enerchi Wellness Press, 2005

Thurman, Robert B., Gerba, C.P. *The Molecular Mechanism of Copper and Silver Ion Disinfection of Bacteria and Viruses.* 1898

Spinella, Marcelio. *Eugenic and Sex Harmony. New York: Elliot Publishing Company, 1936*Haworth Herbal Press, 2005

Yogananda Paramahan. *God Talks with Arjuna The Bhagavad Gita. Los Angeles Self- Realization* Fellowship, 2005

Ra, Un N. A. *Metu Neter, Vol. 1: The Great Oracle Tehuti and the Egyptian System of Spiritual Cultivation.* Bronx, N.Y: Khamit Corp, 1990.

AMEN, RA UN NEFER. *Metu Neter Vol. 2: Anuk Ausar: The Kamitic Initiation System.* 2nd ed. Vol. 2. New York: Kmit Publications, 2009.

Provencio I, Warthen DM (2012). "Melanopsin, the photopigment of intrinsically photosensitive retinal ganglion cells". Wiley Interdisciplinary Reviews: Membrane Transport and Signaling. ***1*** *(2): 228–237.*

I can be reached via my email *addresses*:
cly811@gmail.com or cly811@aol.com

Made in the USA
Monee, IL
08 November 2026

67384876R00100